# THE SKIES ABOVE

# THE SKIES ABOVE

## STORM CLOUDS, BLOOD MOONS, AND OTHER EVERYDAY PHENOMENA

DENNIS MERSEREAU

ILLUSTRATIONS BY REBECCA MATT

MOUNTAINEERS
BOOKS

**MOUNTAINEERS BOOKS** is dedicated
to the exploration, preservation, and enjoyment of
outdoor and wilderness areas.

1001 SW Klickitat Way, Suite 201, Seattle, WA 98134
800-553-4453, www.mountaineersbooks.org

INDELIBLE
EDITIONS

Printed in China
Distributed in the United Kingdom by Cordee, www.cordee.co.uk
25 24 23 22                         1 2 3 4 5

Library of Congress Cataloging-in-Publication data is on file for this
title at https://lccn.loc.gov/2021034865. The ebook record is available at
https://lccn.loc.gov/2021034866.

Mountaineers Books titles may be purchased for corporate,
educational, or other promotional sales, and our authors are available
for a wide range of events. For information on special discounts or
booking an author, contact our customer service at 800-553-4453 or
mbooks@mountaineersbooks.org.

Printed on FSC®-certified materials

MIX
Paper from
responsible sources
FSC® C001701

ISBN (paperback): 978-1-68051-555-8
ISBN (ebook): 978-1-68051-556-5

*An independent nonprofit publisher since 1960*

CONTENTS

To appreciate our atmosphere's beauty, we simply have to go outside. We're surrounded by incredible natural forces that make the sky a celebration of what it means to live on Earth and experience the environment that sustains us. Even hard-core nature lovers and enthusiastic weather geeks sometimes forget to admire the sky's quotidian beauty. We've seen it all before—full moons, foggy mornings, and more sunny days than we can count. But there's a story behind each puffy cloud and raindrop. Even the clearest, calmest day is the product of raging winds and massive storms that lie unseen beyond the horizon.

Everyday weather inspires our plans, our travels, and our dreams. We yearn to travel to the vast expanse of sky and space above us, to the Moon and beyond, always visible but woefully underexplored. Throughout history, the skies have beckoned to brave explorers both on the ground and in the air.

- Galileo Galilei lived under house arrest for daring to suggest that Earth revolved around the Sun rather than the other way around.

- Centuries later, the simple act of predicting a tornado could end an American weather forecaster's career for fear of inciting a panic. Tornado forecasts weren't widely accepted until after 1948, when military meteorologists Ernest J. Fawbush and Robert C. Miller issued the first successful tornado forecast for Tinker Air Force Base in Oklahoma.

■ Apollo 11 carried Neil Armstrong to walk on the Moon just six decades after Orville and Wilbur Wright accomplished the first powered flight. The first astronauts to orbit the Moon were also the first to be awed by the sight of Earth against the emptiness of space. "We came all this way to explore the Moon, and the most important thing is that we discovered the Earth," astronaut William Anders used to say after the Apollo 8 mission. Thanks to weather satellites, everyone on Earth now has the opportunity to see that same view every 15 minutes.

Not only can we gaze upon our planet from far-flung satellites, but we have the ability to explore every part of our world without ever moving an inch. Mapping companies strap 360° cameras to cars and document millions of miles of roads. But for all of the photographs and rocket ships that can take us to destinations most of us can only hope to visit, stepping outside and gazing skyward like every generation before us can still evoke a sense of wonder.

Beyond appreciating its beauty, understanding the basics of how the atmosphere works is vital for staying safe. If awe is the strongest emotion the sky can evoke, fear is a close second. Storm anxiety is a powerful force, especially when flash floods are in the forecast or that a scale-topping hurricane is barreling toward the coast.

The greatest defense against fear is knowledge. After all, most storm anxiety is our imagination filling the void of what we don't know. It's healthy to fear the full power of our natural world. If you're worried about potential storms, it means that you're paying attention. This book provides you with a tool kit indicating what to look for in the skies above. This information, as well as listening for expert warnings and looking for subtle hints in the sky—whether it's the shape of the clouds or that infamous green tinge that paints the base of a heavy thunderstorm—will help you decide whether or not you are truly in danger.

But for all the atmosphere's power, damage, and fury, there are always marvels to be found in how weather works and the astronomical curiosities that lurk in the night sky. If we know how a thunderstorm forms, or why El Niño influences weather conditions continents away, we're more likely to appreciate our surroundings. After reading this book, you'll never miss a chance to take in the wonder of the skies above.

Every raging storm starts with a couple of puffy clouds and a little bit of wind. Understanding the simple concepts that drive our weather can help us appreciate both fierce whirlwinds and placid afternoons. But there's more to our sky than just the weather. Nighttime greets us with an awesome view of a space full of wonders that seem infinitely far away. Whether it's the Moon looming large near the horizon or a dazzling storm of meteors erupting as they hit our atmosphere, the beauty of space and sky is closer than we may realize.

# FRONTS and PRESSURES

Calm days are deceiving. Even a day filled with bright sunshine and a few cooling clouds is influenced by some of the most powerful winds in the atmosphere. Located just a few miles above our heads, these winds routinely roar along at hundreds of miles per hour and are the driving force behind most of the active weather we experience every day.

## Leaving on a Jet Stream

The root of these atmospheric winds is in the changing seasons. There's a tremendous temperature change, or a gradient, between the balmy tropics and the frigid polar climes throughout the year. During the winter months, it's possible to see a temperature difference of as much as 140°F between the lower latitudes and the higher latitudes.

This standing temperature gradient leads to large-scale vertical air circulations such as the Hadley cell. Warm, unstable air in the tropics rises and flows toward colder latitudes, where it cools down and sinks toward the ground. Due to what is known as the Coriolis effect, this sinking air turns parallel to the ground as it descends, leading to narrow belts of wind that encircle both the Northern and Southern hemispheres (see "The Coriolis Effect" sidebar).

These fast-moving belts of wind that flow through the upper atmosphere are called jet streams. They can roar along at more than 200 miles per hour, depending on the season; so fast, in fact, that airplanes flying east across the United States can shave dozens of minutes off their flight times by cruising in the strong tailwind.

Several jet streams can exist at once in each hemisphere. The polar jet stream and the subtropical jet stream are responsible for most weather systems that cross the United States. These upper-level rivers of wind are so powerful that they can pull and push on air located all the way down at the surface.

## Under Pressure

Nature strives for equilibrium. Wind is the atmosphere's cure for an imbalance in air pressure. Air rushes to fill in an area of low pressure, and air flows away to empty out an area of high pressure in the endless pursuit of balance. Twists and turns in the jet stream's shape, as well as regions of faster winds called jet streaks, can cause these powerful upper-level winds to diverge or converge with one another as they enter and exit the jet stream. Divergence induces strong rising motions through the atmosphere nearby, which leads to less air—and lower air pressure—at the surface. Convergence in the upper levels of the atmosphere will cause air to sink rapidly, forcing air to pile up close to the ground and leading to higher air pressure. Earth's rotation comes into play as well, causing winds to circulate in different directions, known as the Coriolis effect.

As a center of low pressure develops and strengthens, air blows in from the surrounding areas to fill in the void left by all the air rising into the upper atmosphere.

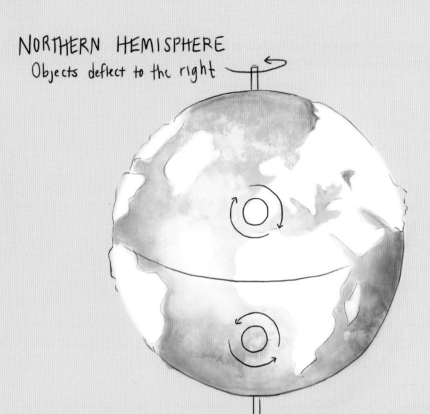

NORTHERN HEMISPHERE
Objects deflect to the right

SOUTHERN HEMISPHERE
Objects deflect to the left

# THE CORIOLIS EFFECT

Earth's rotation plays a key role in which way the wind blows. Air doesn't move in a straight line across the planet's surface, because the planet is always in motion. Earth's rotation causes wind to deflect to the right of its forward motion in the Northern Hemisphere and to the left in the Southern Hemisphere. This deflection is known as the Coriolis effect.

Earth rotates at about 1,000 miles per hour near the equator, the planet's widest point on its axis, with the apparent speed slowing down closer to the poles. Aside from the movement of the Sun, the Moon,

and the stars at night, we don't notice this rotation at all since we're attached to the ground and spinning right along with the planet beneath us.

However, the atmosphere is a fluid above Earth's surface, so it's not firmly attached to the planet as Earth spins on its axis. As a result, wind appears to curve as the ground moves beneath it. The effect isn't all that noticeable for minor weather events, like wind blowing off a lake on a warm day, but it's readily apparent as air spirals around a vast low-pressure system.

The Coriolis effect forces these winds to deflect to the right as they approach the center of the low pressure, which is why winds circulate counterclockwise around low-pressure systems in the Northern Hemisphere. Low-pressure systems will continue to strengthen as long as there's divergence in the atmosphere above; as soon as the jet stream weakens or smooths out, the rising motions will slow down and air will begin filling in the low-pressure system. Stormy weather tends to subside and conditions improve when low-pressure systems weaken and dissipate.

High-pressure systems can persist as long as convergence pushes air from the upper levels toward the surface. If this sinking motion weakens, the area of high pressure will start to equalize with its surroundings. The Coriolis effect causes air to flow in the opposite direction for high-pressure systems: the buildup of air at the surface causes winds to flow away from the high-pressure systems, curving and rotating clockwise in the Northern Hemisphere.

These centers of low pressure and high pressure drive so much of the weather we see on a daily basis. A high-pressure system is relatively boring if you're into exciting weather. Since air dries out and warms up as it descends, conditions beneath a ridge of high pressure are typically pretty nice. But too much of a nice thing can be a problem. Strong highs during the summer can lead to brutal heat waves. If a stubborn weather pattern causes a broad area of high pressure to linger for weeks or even months at a time, that region can slip into drought. Most highs are short-lived, though, so they're usually good for a couple of pleasant days between storm systems.

Low-pressure systems, on the other hand, are the culprit behind just about every memorable weather event you can think of. All low-pressure systems are called cyclones. There are several different types of cyclones. The two most common we deal with on a regular basis are tropical cyclones—including hurricanes—and extra-tropical cyclones.

The vast majority of lows are extratropical cyclones. These low-pressure systems can move massive amounts of air around the world. A cold gust of wind on a winter's

> There is really no such thing as bad weather, only different kinds of good weather.
>
> —JOHN RUSKIN, quoted in *The Use of Life*

day likely started hundreds or thousands of miles away over the ice fields of the Arctic. The leading edges of these air masses on the move are called fronts. Since so much of our daily weather is driven by the intricate movements of these air masses, tracking the position and movement of fronts is a big part of weather forecasting.

## Cold Fronts

Cold fronts are a force to reckon with. It's exhilarating to stand outside as a strong cold front passes through and the temperature plummets in just a matter of minutes. Cold fronts, which are often displayed on weather maps as blue lines with triangles pointing in the direction of movement, represent the most dramatic air-mass changes as low-pressure systems pass by.

Cold air is dense, so it wants to hug the surface any chance it gets. If we could take a cross section of a cold front as it rolls across a town, we'd see a bubble of cold air rolling over the ground. This dense air crawls along the surface as it moves into warmer air ahead of it, scooping that warm air high into the atmosphere and leaving cold air in its wake.

That kind of lift is very efficient at creating thunderstorms. The swift-rising motions along the leading edge of a cold front can lead to intense lines of storms that can traverse hundreds of miles. These storms are often innocuous, but if there's enough warm, rising air (or instability) and the upper-level winds are strong enough, these storms can produce destructive wind gusts and tornadoes.

It's not just the weather along a cold front that makes them so fascinating—it's also the conditions that follow behind them. A vigorous cold front can send temperatures from balmy to frigid in record time. The most dramatic temperature drop ever recorded in the United States occurred when a cold front swept through Rapid City, South Dakota, on January 10, 1911, dropping the temperature from 55°F to 8°F in just 15 minutes.

## Warm Fronts

Warm fronts are less dramatic than their colder counterparts, but they're impressive nonetheless. A solid warm front can raise temperatures as quickly as a cold front can drop them. It's not uncommon in the southeastern United States for a chilly winter's day to give way to an afternoon balmy enough to need the air conditioning as a warm front slides through the area.

# COLD FRONT
DIRECTION OF FRONT →

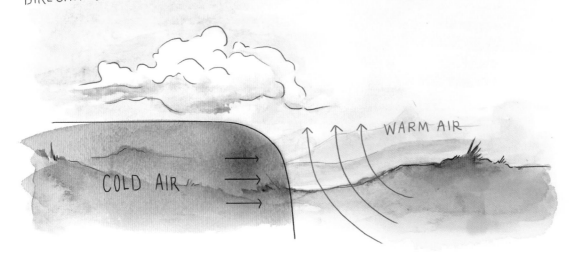

WARM AIR

COLD AIR

# WARM FRONT
DIRECTION OF FRONT →

WARM AIR

COLD AIR

A cross section of a warm front looks like the opposite of a cross section of a cold front, since warmer air is less dense and rises over the top of colder air at the surface. The leading edge of a warm front passes thousands of feet above a particular location long before the warm front's arrival is felt at the surface. The surface front is usually displayed on weather maps as a red line adorned with semicircles pointing in the direction of movement.

Conditions ahead of a warm front are often gray and rainy because the warm air overruns the cold air beneath it, sometimes transitioning over to a heavy drizzle or thick fog once the region where the warm front has reached the surface finally passes through. Since the air behind a warm front is often humid and unstable, the warm air mass is a prime area for severe thunderstorms to develop. The warm front itself can actually serve as a focal point for storms to produce dangerous tornadoes.

## Stationary Fronts

Sometimes, warm and cold air masses are in no hurry to clash with each other. Weak or recently dissipated low-pressure systems can result in two different air masses stuck next to each other. The boundary between adjacent cold and warm air masses is called a stationary front. You can experience a quick change in weather conditions if you drive across one of these sluggish boundaries, going from cold and snowy to warm and rainy in just a few miles.

## Occluded Fronts

The fourth type of front, an occluded front, occurs when a cold front catches up with and overtakes a warm front. This complicated dance of intertwining air masses occurs after a low-pressure system achieves its peak intensity and begins to wind down. An occlusion usually starts the process of choking off a low-pressure center and causing the system to fill in and eventually dissipate. Weather conditions calm down and clear out as a low-pressure system weakens and falls apart. Sometimes, the remnants of one low-pressure system can serve as a focal point for the development of another as the jet stream meanders downwind.

# RAIN, SNOW, and PRECIPITATION

**W**e have a conflicted relationship with precipitation. Depending on its form and quantity, rain and snow can be a fantastic relief or a worrisome burden. Precipitation is essential to human life, and yet it's wiped out entire communities in a flash. Whether it's a raging thunderstorm or a fleeting snow shower, precipitation represents the awe-inspiring power of our atmosphere to create so much from so little. This process is such a basic, foundational part of our everyday lives that we barely pay it any mind.

## Rain, Rain, Go Away

Rain is as diverse as the clouds that produce it. Even though it all splashes on your window the same, it is phenomenal how many different ways our atmosphere can produce rain. There's stratiform rain, convective rain, tropical rain, and rain that forms from snowflakes that melt before they ever reach the ground. No matter how rain forms, though, it all starts with water vapor.

Precipitation requires lift in order to form a cloud. This external force can come in the form of cold fronts, sea breezes, air rising through convection, or the large-scale ascent we would see associated with a sprawling low-pressure system. Rising water vapor, an invisible gas that results from evaporation, condenses into clouds made up of microscopic water droplets. These water droplets require a nucleus—usually specks of dust or pollution suspended in the atmosphere—around which they can begin to coalesce into a raindrop. It takes millions of these microscopic water droplets to create raindrops.

A single cloud can weigh millions of pounds. It's hard to wrap the mind around how heavy clouds are as we watch these billowing formations float effortlessly in the skies above. Each of those astonishing formations is made up of countless individual water droplets that can combine into a torrential downpour.

The type of lift involved in creating a raindrop determines the extent and intensity of the rain that falls. The most common type of rain is stratiform precipitation, which forms from a widespread region of air slowly rising as a result of a passing trough or nearby low-pressure system. Since the lift is low and slow, stratiform rain falls at a gentle, steady pace—the kind that's ideal to listen to as you drift off to sleep. You can experience this kind of rain just about anywhere in the world that enjoys seasonal weather changes or a cool, coastal climate (such as Seattle, Washington). Chicago putters through plenty of rainy days, but down in Tahiti, where conditions rarely diverge from one day to the next, it's hard to come by more than a passing shower or storm.

However, rain isn't always peaceful. Intense thunderstorm updrafts fueled by convection, or warm air rising through colder air above it, can produce huge raindrops and a phenomenal amount of rain in a short period of time. Sometimes it's an absolute downpour that seems more like a movie special effect than real water gushing from the sky. Regions around the Gulf of Mexico are famous for their near-daily deluges in the warmer months. Mobile, Alabama, received more than three inches of rain in just over 20 minutes on April 29, 2014, as a stationary thunderstorm raged over the city. Just a few miles to the east, Pensacola, Florida, measured nearly six inches of rain in a single hour that same night, and some neighborhoods nearby recorded almost two feet of rain within 24 hours.

Tropical cyclones produce bouts of blinding rain that can last many hours longer than your typical summertime thunderstorm. A tropical cyclone, a type of low-pressure system that includes tropical storms and hurricanes, produces rainfall through the warm rain process. Rather than raindrops growing slowly in layers, the warm rain process involves cloud droplets directly colliding with one another. These microscopic collisions within the clouds churn out tons of tiny raindrops. This process makes tropical cyclones highly efficient rainstorms, which is why these systems are such a grave flash flood threat wherever they make landfall.

It's not the air alone that plays a role in creating precipitation. The ground itself can influence the formation of the precipitation to feed its own nourishment. This effect is most profound through orographic lift, or moist winds sweeping up one side of a mountain range. Mountains are really effective at wringing out the moisture from air as it flows up the slopes. Air cools off as it rises up the side of a mountain, allowing water vapor in the air to condense into clouds that produce heavy rain.

Orographic lift is responsible for the incredible rain shadow effect in places like the state of Washington. As storm systems come ashore from the Pacific Ocean, they can bring many feet of precipitation in a normal year to communities in western Washington. As the moist winds from the Pacific continue eastward and rise over the western Cascade Mountains, they release most of their moisture before flowing down the eastern slopes. Winds descending down the eastern slopes of the Cascades dry out and warm up as they reach lower elevations, leaving these areas exceptionally dry even as communities a few dozen miles away reside in a temperate rain forest climate.

> The sky was dark and gloomy, the air was damp and raw, the streets were wet and sloppy … the rain came slowly and doggedly down, as if it had not even the spirit to pour.
>
> —CHARLES DICKENS, *The Pickwick Papers*

# HURRICANE HARVEY'S FLOODING

The most rain ever recorded during a landfalling hurricane in the United States occurred during Hurricane Harvey in 2017. The storm made landfall in Texas and stalled over the region for several days. The hurricane, which quickly weakened to a tropical storm, dumped copious amounts of rain as it meandered along the northwestern Gulf Coast. Harvey produced a record 60.58 inches—that's more than five feet!—of rain in Nederland, Texas, between August 25 and August 29. It produced so much rain in such a short period of time that widespread flash flooding required the rescue of thousands of people across southeastern Texas and southwestern Louisiana.

Cities themselves can even influence weather patterns around them. Cities are hot. Thousands of buildings, endless roads and parking lots, and a 24-hour barrage of vehicle traffic raise the temperature of dense cities and suburbs significantly higher than their less urbanized surroundings. This temperature difference is known as the urban heat island effect. A 2014 study by Climate Central found that the average temperature in downtown Las Vegas, Nevada, was 7.3°F warmer than the surrounding areas during the summer. That's no small matter when Las Vegas's average high temperature climbs above 100°F for three months of the year. Not only does the urban heat island effect raise temperatures, but the warm air generated by an urban area can induce convection that triggers heavy rain showers and even thunderstorms over the city that influenced their creation. These thunderstorms are especially noticeable in cities in the southern United States, such as Atlanta, Georgia, where high heat is often coupled with high humidity.

## Let It Snow

Magical things happen when the temperature dips below freezing. The world seems to slow down, including nature. And the processes that go into precipitation become vastly more complicated. Precipitation is more than just rain and snow. There's a whole spectrum of wintry precipitation that can form, depending on precise temperatures and the moisture in the air.

the snow doesn't give a soft white damn Whom it touches.

—e.e. cummings, *ViVa*

Snow forms through a similar process as rain. If temperatures are far enough below freezing at cloud level, supercooled water droplets— water that remains liquid when it cools below freezing—can latch onto an impurity in the clouds and instantly freeze into an ice crystal. This process creates snowflakes that eventually fall to the ground when they're too heavy for lift to keep them suspended in the cloud.

While we're familiar with the classic shape of a snowflake, called a dendrite, there is a whole slew of shapes that snowflakes can take depending on how cold it is and how much moisture is in the air. Snowflakes that form when it's very cold or relatively dry are smaller and have a low moisture content. This is the type of powder that folks love to play in when they go to ski resorts. Snowflakes that form in a high-moisture environment,

# NO TWO SNOWFLAKES ARE ALIKE—TRUE?

We learn at a young age that no two snowflakes are the same. It's one of those sayings that almost sounds too good to be true. Physicists largely agree that complex snowflakes—dendrites with intricate branches and scratch-like markings throughout the ice crystals—can be arranged in so many different ways that the odds of two snowflakes forming to be exactly the same, with identical features and markings, is exceptionally small. You could spend a lifetime staring at countless snowflakes through a microscope and never come across two flakes that are exactly alike.

when temperatures are close to freezing, are on the chunky side. These are the giant aggregate flakes that seem to glop out of the sky and hit the ground with a muted splash.

The atmosphere is a dynamic fluid that's rarely uniform from top to bottom. There are always small layers of warmer and colder air creeping around at some altitude above the surface. Snow is both resilient and fragile at the same time. If temperatures remain below freezing from the bottom of the clouds all the way to the ground, all precipitation will fall as snow. But if a little bit of warm air nudges its way between the clouds and the ground, the snowflakes will begin to melt. The amount of melting determines what type of precipitation reaches the surface. For example, snowflakes can survive temperatures as high as 40°F if the layer of warm air is very shallow and right near the ground.

If snowflakes melt only a little bit before they enter freezing air near the ground, the partially-melted snowflakes will refreeze into ice pellets. These ice pellets are commonly called sleet. Sleet lands on the ground with a characteristic "tink" that can sound like a sizzling pan when it's heavy. Sleet looks just like snow when it's on the ground, but it can freeze into solid ice very quickly. Frozen sleet on the ground is extremely difficult to remove, making it hard to walk or drive in the days after a sleet storm.

# TYPES OF SNOWFLAKES

The shape of a snowflake depends on the temperature and moisture within the clouds. Temperatures just around the freezing point create snowflakes in the shape of thin plates, often decorated by intricate markings that form as the snowflake grows. Bitterly cold temperatures create snowflakes in the shape of columns.

The most common snowflakes are dendrites, named for the tree-like branching that defines their iconic shapes. Dendrites appear most often because the conditions needed for their formation, which include high moisture and temperatures between 0°F and 10°F, occur frequently, allowing supercooled water droplets to freeze efficiently and develop into beautiful and unique flakes.

Snowflakes that completely melt in a layer of warm air before reaching the surface become pure liquid raindrops as they approach the ground. Since these liquid raindrops contain no ice crystals around which to freeze, the water itself can become supercooled. These supercooled raindrops, called freezing rain, can freeze on contact with any exposed surface, instantly creating a glaze of ice on whatever they hit. The most destructive ice storms can leave a crust more than two inches thick, which is heavy enough to snap the hardiest trees and power lines.

# A snow day literally and figuratively falls from the sky, unbidden, and seems like a thing of wonder.

—SUSAN ORLEAN, "Snow Day," *New Yorker*

# NAME THAT CLOUD

Our sky is in constant motion. We can't always see the wind as it blows, but we can view clouds as they drift across the sky. Time-lapse footage from the International Space Station shows that clouds in the atmosphere roil like a boiling pot of water. These formations can change weather in a flash, but they often move so slowly above us that we hardly notice their endless theatrics.

The variety of clouds in the sky depends on prevailing weather patterns moving across an area. A high-pressure system overhead may allow only a few high-flying cirrus clouds or some puffy fair-weather cumulus to dot the sky. A dynamic system, such as a winter storm or hurricane, can produce in just a couple of hours an impressive display of a dozen or more types of clouds woven like an intricate tapestry throughout the sky.

Each cloud is classified by both a group and a type. However, even the most astute weather observer can find it difficult to keep track of all the cloud types and accurately identify what's floating around in the sky on any given day. The easiest way to keep track of clouds is to remember that the three major groups are stratus clouds (smooth), cumulus clouds (puffy), and cirrus clouds (wispy).

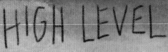

# HIGH LEVEL
### ABOVE 18,000 FT

CIRROCUMULUS

CIRRUS

CIRROSTRATUS

# MID LEVEL
### 6,500 – 18,000 FT

ALTOSTRATUS

ALTOCUMULUS

STRATOCUMULUS

# LOW LEVEL
### 0 – 6,500 FT

CUMULUS

NIMBOSTRATUS

CUMULONIMBUS

STRATUS

# Stratus

**Stratus clouds** are the product of air rising slowly, a gentle simmer instead of a rolling boil. Passing weather systems cause these clouds to form in striated layers or decks, often appearing to cover the sky as far as the eye can see. Casting a dour pall, they don't always produce precipitation. A slow, gentle lift along a warm front or near the center of a robust low-pressure system can lead to a deck of stratus so expansive it can cover more than half of the United States at once.

Stratus clouds that produce precipitation are called nimbostratus clouds. Nimbostratus decks lend a gloomy, imposing presence to the sky when they loom overhead. These dark, almost featureless clouds hang low to the ground—so low, in fact, that they can obscure the tops of tall buildings in cities.

GUIDE TO SPOTTING: If the sky is uniformly dark and gloomy and it's lightly raining or drizzling, it's probably nimbostratus.

**Altostratus clouds** are mid-level clouds that form a few thousand feet above the surface. The hallmark of altostratus clouds is a watery sun that shines through the clouds like a blurry shimmer on a murky pond. Cirrostratus clouds form in the upper levels of the atmosphere. These clouds lend a milky appearance to the high sky, sometimes leading to a vivid halo around the Moon or bright and colorful reflections around the Sun known as sundogs. If you ever hear a a weather report mention "fair skies," it means that cirrostratus clouds are present.

GUIDE TO SPOTTING: Altostratus and cirrostratus clouds are high and thin enough that they don't noticeably affect weather on the ground.

# Cumulus

While stratus clouds form from air that rises slowly, conditions grow markedly more interesting as the rising air picks up speed. The rapid lift associated with convection, or along cold fronts or sea breezes, can instigate the formation of **cumulus clouds**. While most cloud groups and types are classified by their altitude, cumulus clouds can stretch from the lower levels to the upper levels of the atmosphere.

Cumulus clouds are convective clouds. Convection occurs when warm air rises through cooler air above it. This pocket of ascending air slowly cools off as it gains altitude. Cooler air can hold less water vapor than warmer air, so as the air's temperature decreases, its humidity increases. Once the rising air reaches 100 percent humidity, the water vapor in the rising air condenses and forms a cumulus cloud. Eventually, the temperature of the rising air will equalize with the environment around it, allowing the cool air to sink back toward the ground. This convective loop of rising and sinking air gives the classic cumulus cloud its domed shape.

GUIDE TO SPOTTING: If the clouds look like cotton balls against a blue sky, they're cumulus.

Most convection is benign, which leads to fair-weather clouds that can blot out the sunshine for a few minutes. Sometimes, though, the temperature difference between lower levels and upper levels is so great that this unstable air can blast skyward at more than 100 miles per hour. The amount of instability in the atmosphere determines if cumulus clouds will grow into something more than just billowing decorations.

If these columns of rising air, called updrafts, rise quickly enough, the cumulus cloud can turn into a **cumulonimbus cloud** that produces heavy rain, heavy snow, hail, and thunder and lightning. While often weak and short-lived, some produce flooding rains, destructive hail, damaging wind gusts, and even tornadoes.

When the updraft that feeds a towering cumulonimbus rises as high as it can, the air abruptly stops rising and begins to spread out, dragging the cloud along with it. This dramatic plate-like formation is known as an anvil, since the shape of the cumulonimbus and its spreading top resemble a blacksmith's anvil. Since only well-developed thunderstorms to grow miles into the sky, anvil clouds are a decent warning sign that you're looking at a potent thunderstorm.

**GUIDE TO SPOTTING:** If a hulking cumulus cloud towers above the horizon like a mountain, it's very likely a cumulonimbus cloud.

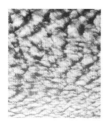

While cumulus clouds often stretch from the bottom to the top of the atmosphere, some types of cumulus do form in the middle and upper altitudes. **Altocumulus clouds** are mid-level formations that often occur in large fields, sometimes filling the sky. A deck of altocumulus can appear as a "mackerel sky," covered by rows of tiny, individual clouds that resemble fish scales.

**GUIDE TO SPOTTING:** Altocumulus clouds are about the size of your thumbnail when looking at your hand held against the sky.

**Cirrocumulus clouds** are high-level cumulus clouds that look similar to altocumulus clouds. The most memorable of these formations are the result of turbulent winds blowing over mountains, appearing in the sky as if they were tight ripples on the surface of a still pond.

**GUIDE TO SPOTTING:** Cirrocumulus clouds look like pebbles dotting the top of the sky.

## Cirrus

**Cirrus clouds** are feathery formations that glide along in prevailing winds near the top of the atmosphere, usually around the cruising altitude for jet aircraft. Cirrus are typically made up of ice crystals due to the extremely cold temperatures in the environment around them, which can dip below -50°F. The arrival of cirrus clouds doesn't necessarily foretell the approach of any particular storm system; these clouds are as common during calm weather as they are ahead of a powerful hurricane.

**GUIDE TO SPOTTING:** Cirrus clouds look like floating feathers.

# HUMANS CAN MAKE CLOUDS TOO

Clouds don't always form from natural means. Human activity influences the formation of clouds so commonly that we seldom notice. For instance, high-flying jet airplanes leave thin condensation trails in their wake. These thin, wispy clouds, known as contrails, are the result of water vapor in extremely hot jet exhaust instantly condensing into clouds in the frigid air of the upper atmosphere. It's the same basic principle behind why you can see your breath on a cold morning, just on a much larger scale. Depending on moisture levels and winds, these contrails can dissipate immediately or linger for many hours, sometimes even spreading across the sky as a thin deck of cirrostratus.

Condensed jet exhaust isn't the only cloud an airplane can produce. Pilots can attach canisters full of paraffin oil to small aircraft to write smoke messages in the sky. Smoke from paraffin oil heated by engine exhaust leaves a clear pattern in the sky behind the path of the aircraft. A pilot can start and stop the smoke at will, allowing them to draw letters and shapes with ease. Since the smoke from this oil mixture lingers and takes a while to dissolve, on a clear day with calm winds, skywriting may remain visible from miles around for up to half an hour.

Power plants and factories that vent extensive amounts of steam can trigger the formation of cumulus clouds near the ground. These clouds can extend thousands of feet into the air above the steam stacks at these plants. Communities near nuclear plants have even experienced rain and snow showers from these steam-induced clouds if there's a great enough temperature difference between the lower and middle levels of the atmosphere.

# Unusual Clouds

Some types of clouds don't fall neatly into the three big groups. **Actinoform clouds** are unusual formations that we didn't know existed until weather satellites launched in the 1960s. These over-the-ocean maritime clouds don't look like much from the surface, but from above they reveal complex spirals and fractal patterns that seem to wiggle and spin above the ocean surface. Meteorologists still aren't quite sure how these cloud formations develop or why they favor colder waters like the southeastern Pacific Ocean.

**Mammatus** are high-level clouds that typically form along the anvil of a cumulonimbus cloud. It's no coincidence that these clouds are named after mammary organs, as they are deep, bubbly, bulging pockets of clouds that dip below a cloud deck as a result of extreme turbulence in the atmosphere. A well-formed field of mammatus clouds can look otherworldly, especially if these formations manage to catch the colors of a vivid sunset. Pilots use these clouds as a warning to turn away and avoid potentially dangerous turbulence.

Mountain ranges are tall enough to have a profound impact on the clouds around them. One of the most striking clouds formed by the mountains is **lenticular clouds**. These lens-shaped clouds form when moist winds cool as they ascend a mountain slope, condensing their water vapor as they approach and round the peak of the mountain. The wind warms up and dries out as it falls back down the other side of the mountain, leaving behind a smooth lenticular cloud. The effect is most pronounced on a tall mountain during a stiff wind. Sometimes multiple lenticular clouds form above a mountain, stacked on top of one another like pancakes.

# PREDICTING the WEATHER

**A**ccurate weather forecasting is a marvel of human ingenuity. Meteorologists are now able to predict the weather with impressive precision up to a week in advance, which would have been an unimaginable feat just a few generations ago. The rapid advance of technology, knowledge, and mass media almost allow us to take weather forecasts for granted. For as common as they are, these daily reports are a lifesaving testament to the power of science—and the inescapable force of the world around us.

## Looking Back at Forecasting

A weather forecast is only as accurate as the tools available to the forecaster. Before the mid-1900s, forecasters had to rely on basic tools and folklore to predict coming conditions. Most of the old methods were based on deductions passed down from one generation to the next.

Barometers, which measure atmospheric pressure, were a mainstay of pre-computer weather forecasting, because changes in air pressure are a sign of changing weather patterns.

Other methods looked to the natural world for clues. The leaves on a deciduous tree turning inside out could signal thunderstorms on a summer's day because the supple green leaves are sensitive to the humidity. If the leaves bend to show their lighter underbelly during a gust of wind on a warm afternoon, it could mean that a torrent is going to bubble up in the humid air.

Thanks to the rapid advance of technology through the last century, you don't have to rely entirely on Uncle Bob's achy back or the twisting leaves of an oak tree for clues about tomorrow's weather. You can look at satellites, weather radar, and computer models to predict the weather tomorrow and in the coming week.

## Satellites

For a weather geek, one of modern life's little pleasures is reviewing a vivid satellite image. It's truly a wonder that we're able to get a real-time look at our planet from space. The practice of tracking weather systems using satellites began in 1960 when the United States launched TIROS-1 into low orbit. The satellite consisted of a wide-angle television camera pointed toward Earth, transmitting its pictures back to antennas on the ground. Today's weather satellites are significantly more advanced, using sensors to collect data and imagery in stunning resolution.

The predominant family of weather satellites in the Western Hemisphere is the Geostationary Operational Environmental Satellite (GOES) program. Operated by the US National Aeronautics and Space Administration (NASA) and the US National Oceanic and Atmospheric Administration (NOAA), GOES satellites are launched in geostationary orbit so the satellite is always fixed in a single position above Earth's surface. A satellite achieves this feat by orbiting at about 22,000 miles above the equator, putting its orbital speed at precisely the same speed at which Earth rotates.

Geostationary orbit is important because it provides the satellite the exact same view of the planet for the satellite's entire life span, allowing for a clean image of the skies from above without interruption.

The first satellite in the GOES family was launched in 1975. As of the last few years, there have been 17 GOES satellites launched into orbit, with many more planned in the years and decades to come. Each new series of satellites is more advanced than the one it replaced. The fourth series of GOES satellites, first launched in 2016, can take detailed snapshots of the atmosphere every 30 seconds, with a resolution so fine that the satellite can detect small wildfires the moment they flare up. That's a tremendous improvement from the first GOES satellite, which took one fixed picture of the hemisphere every 20 minutes.

A weather satellite works by detecting visible and infrared wavelengths reflected by the planet below. A high, cold cloud in the atmosphere radiates a different wavelength than the hot ground of a desert. Satellites transmit their findings back to computers on the ground, where software converts the raw data into the visible images we see on the nightly weather report. Not only can we see imagery of clouds and water vapor, but the latest satellites can spot individual flashes of lightning and keep track of solar storms that could threaten communication networks and the health and safety of astronauts in orbit.

## Weather Radar

In addition to information gleaned from satellites, weather radar can provide valuable minutes to hours of advance warning to prepare for an approaching storm. Data collected by weather radar can tell us the location, intensity, and type of precipitation falling from the clouds.

Weather radar works by sending out a powerful beam of microwave energy from a rotating dish nestled within a protective fiberglass dome that sits atop a tall tower. This microwave radiation zips through the atmosphere and hits any objects in its path—rain, snow, hail, flying cows—and that reflected energy returns to the radar dish. The amount of time it takes for the radiation to return to the dish, as well as the intensity of the radiation reflected back to the dish, allows computer software to calculate the coordinates and intensity of whatever is in the sky.

Modern radar equipment can even use the Doppler effect to measure how quickly objects are moving toward or away from the radar tower. The Doppler effect follows

# THE AMAZING ACCURACY OF WEATHER PREDICTIONS

Meteorologists have a thankless job. They predict the future every day and get it right with astounding frequency, but folks only ever remember when they miss a big storm or predict a warm day that turns out to be chilly. Despite the long-running joke about meteorologists getting it wrong half the time and still getting paid, these scientists more than earn their keep.

A weather forecast on Tuesday can predict Saturday's temperatures with a level of accuracy meteorologists 30 years ago could only hope for. Today's weather models can point meteorologists toward flash floods hours before they unfold. The US National Hurricane Center's three-day hurricane-tracking forecasts in 2020 were as good as a one-day tracking forecast in 1995. It may not seem like much, but that's at least two extra days for coastal residents to prepare for a dangerous storm.

the same principle that makes an ambulance's siren sound louder and faster when the ambulance is racing toward us than it does when the ambulance is driving away. Observing this change in the radar beam's frequency lets us measure wind speed and direction. The invention of Doppler weather radar was a tremendous public safety success, providing up to an hour's warning ahead of hazards such as tornadoes and damaging wind gusts caused by thunderstorms.

Most countries that routinely experience severe thunderstorms and tropical cyclones have a lifesaving network of weather radar sites monitoring their skies. The United States is protected by a network of more than 100 weather radar sites spread across the country and its territories. Canada upgraded its radar network in the late 2010s to improve coverage across the vast country's population centers, and many other countries around the world have built radar networks of their own.

## Weather Stations
While we've outgrown our dependence on old folklore to predict the weather, there's still untold value in going back to the basics to keep tabs on current conditions. Weather stations are a vital part of both weather monitoring and forecasting. Reliable weather records stretch back more than a century in most communities across the United States, building a solid baseline for the kind of weather conditions a region normally experiences and giving us an idea of the kind of weather conditions a region may expect in the future.

Modern weather stations include a thermometer to measure temperature, a hygrometer to track moisture, an anemometer to record wind speed, a barometer to follow air pressure, rain gauges, and a bevy of other devices to collect data on everything from visibility to solar radiation.

Most weather observations are collected at airports around the world, but the growth of internet-connected devices has rapidly expanded the use of home weather stations. Certain companies allow users to publish weather conditions right from their backyard, which lets folks get real-time weather conditions just down the street.

The most common use for data collected by weather stations is simply to check the temperature or see how windy it is outside. But real-time updates can alert meteorologists about sudden changes in weather, such as dangerous wind gusts in approaching thunderstorms, so they can issue warnings to people in harm's way.

# Weather Balloons

A weather station gives us fantastic data about what's going on at ground level, but this is only a tiny sliver of the entire atmosphere. How do we find out what's going on miles above our heads? It's simple—we just send floating weather stations into the sky!

Weather balloons, while often the butt of jokes about UFO sightings, are an indispensable part of weather forecasting around the world. The practice of collecting weather data using balloons came about in the early 1900s and proliferated with the development of radio and satellite technology later in the century. Meteorologists release hundreds of thousands of weather balloons every year from stations around the world, including more than 125,000 from North America alone.

The process is both simple and a technological marvel. An instrumentation box known as a radiosonde is attached by a long string to a large latex or synthetic-rubber balloon. Observers fill the balloons with either hydrogen or helium and release them into the sky. Sometimes the balloon blows into a tree, but most of the time they rocket skyward and disappear in a matter of seconds. Their flights typically last two hours.

A parachute attached to the radiosonde ensures that the box doesn't hurt anyone when it returns to the ground. Most of the balloons fall into the ocean or uninhabited lands—creating an unfortunate littering problem—but occasionally someone will find one in their backyard and can return the equipment to the agency responsible for launching it.

These instrument boxes collect data about wind, temperature, moisture, and air pressure all the way to the top of the atmosphere. Global Positioning System (GPS) devices within the instrument box track the balloon's position, speed, and altitude as it races through the atmosphere, telling us which way and how fast the wind is blowing.

Data collected by radiosondes is critical to weather forecasting. Analyzing the temperature profile of the atmosphere helps meteorologists determine if precipitation will fall as rain, snow, or ice. A plot of temperature, moisture, and winds can help a forecaster diagnose the threat posed by severe thunderstorms and tornadoes.

# Computer Models

All the information gathered by those hundreds of daily weather balloon launches gets fed into computer models. Incredibly creative scientists developed computer models to simulate the atmosphere, the land, and the oceans using mathematical and physical equations. These weather models were simple at first, simulating only some aspects of the atmosphere and predicting the weather only a day or so in advance, but the fine resolution, extended range, and high accuracy of today's weather models are a tremendous boon to our ability to keep tabs on what lies ahead.

Weather models use a blend of climatological statistics, current weather information, and basic principles of physics and meteorology to arrive at their final product. Data from weather balloons, ground-based weather stations, radar, and satellites gets ingested into weather models.

The output from a weather model can predict just about every aspect of the atmosphere, from jet streams to temperatures and even the trajectory of wildfire smoke and volcanic ash. Global weather models, such as those run by NOAA and the European Centre for Medium-Range Weather Forecasts (ECMWF), simulate large-scale weather systems as they circle the world. These models are great at simulating the jet stream and big storms like nor'easters and hurricanes. Small-scale models—programs that cover a country rather than the whole world—can run at a much higher resolution, sometimes even spotting the location and intensity of an individual thunderstorm hours before it bubbles up.

# PREDICTING RAIN WITH ACHY JOINTS

If you've ever injured your knee or back, you probably know when it's going to rain before the first drop falls. Our joints are full of synovial fluid, a slippery substance that acts like a lubricant allowing bones to slide past each other without injury. This fluid expands and contracts as air pressure falls and rises. Falling air pressure, like you'd experience ahead of and during a rainstorm, causes the joint fluid and surrounding tissues to expand, resulting in pain that's even worse for folks with existing injuries and arthritis. The discomfort is exacerbated during cold weather, when the synovial fluids thicken a bit and the tissues surrounding the joints tighten up—which supports the decision of every retiree who fled the harsh northern winters for the warm, high-pressure relief of states like Arizona and Florida.

## Forecasting as an Art

Predicting the weather is as much an art as it is a science. Deciphering weather models and analyzing scores of data requires a heap of scientific knowledge and technical know-how, but the process also requires experience.

Meteorologists refer to computer models as "guidance" because they're intended to aid a forecaster in predicting the weather. Algorithms are great, but they're not perfect. All models have their own weaknesses and biases that can yield flawed results. Trained forecasters will use multiple models, along with their experience and knowledge, to come up with the best solution.

Sometimes a weather model spits out bad guidance, or that weather balloon pops, or the weather radar gets struck by lightning and goes offline. When circumstances like these arise, meteorologists can rely on their knowledge and experience—and on rare occasions, even a hunch!—to figure out what the weather is going to do next. And, more often than not, a humble and experienced forecaster will get it right.

Scientists are always working toward improving their weather forecasts, but they may never achieve perfect accuracy. The atmosphere is too vast and complicated to predict every little wave and wind shift. But forecasters can take steps toward improving accuracy and precision. Someday soon, longer-range weather forecasts—perhaps a week or more in advance—will be even more reliable, throwing us fewer curveballs and helping people better prepare for hazards to come.

# WEATHER vs. CLIMATE

Nine of the world's 10 warmest years on record occurred between 2010 and 2020. Every year but 2011 and 2012 landed on that dubious top-10 list, with 2016 and 2020 each marking global temperature records when they ended nearly 1.76°F above normal. The hard data confirms what we already know: the climate is changing, which affects the weather.

Weather and climate are two distinct fields, yet they are inextricably linked to each other. Keeping tabs on the weather requires knowledge of what the weather normally does in order to figure out what the weather might do in the future. It's important to understand both weather and climate to develop a deeper appreciation for our atmosphere.

Weather is the state of the atmosphere. It's the temperature and the humidity, a raging jet stream and a gentle breeze, a fine drizzle and a mammoth hailstone. When we talk about the weather, we're discussing what the atmosphere is doing right now or what it will do over the next couple of days. A weather *report* might reveal that it's 52°F with a light wind that feels chilly beneath a thick cloud coverage. A weather *forecast* might show that tornadoes are possible tomorrow or that a hurricane could approach the coast in five or six days.

Just about every community in the United States has an old joke that goes "if you don't like the weather, just wait five minutes." Weather conditions can fluctuate wildly from one day to the next, and sometimes lurch between extremes over the course of a single day. During an active winter's day, it's not uncommon for Miami to bask in temperatures warm enough to break a sweat while communities a few hundred miles away shiver in subfreezing temperatures.

# Climate is what we expect, weather is what we get.

—ROBERT HEINLEIN,
*Time Enough for Love*

Given the extreme temperature gradients that exist between the equator and the poles, weather patterns grow more active with distance from the lower latitudes. Daily weather conditions are more reliable in tropical latitudes, where just about every day features high heat, sultry humidity, and a chance for heavy rain or thunderstorms. Farther away from the equator, extreme temperature changes lead to extreme pressure changes, a turbulence that churns the atmosphere in thrilling and terrifying ways.

Though it may not seem like it at times, the wild swings in weather conditions from day to day average out to a relatively smooth pattern throughout the year. Temperatures ebb and flow as the seasons change. Precipitation amounts are relatively predictable each month, often spiking late in the summer and falling off during the heart of winter. The average of weather conditions over a period of weeks, months, and years is known as climate.

Climatology tells us that a 52°F afternoon is 10 degrees below normal for a late spring day. A history of reported tornadoes over the last 50 years reveals that tornadoes are more common in the Great Plains than the Mid-Atlantic states during the month of May. We can adequately prepare for hurricane season in advance because climatology reveals that most hurricanes in the Atlantic Ocean form between June 1 and November 30. Climatic data is simply the history of the weather, and just like human history, knowing and understanding what's happened beforehand gives us a solid perspective on what to expect going forward.

Meteorologists often succinctly describe the difference between weather and climate as the difference between a person's mood and their personality. It's possible to have a bad day and act irascible toward people you know and love, even if you're normally chipper and affable. The atmosphere works the same way. It's possible for a

# DEVASTATING WILDFIRES

Longer-lasting and more intense droughts can lead to more than water woes and decimated crops. Lengthy periods without meaningful precipitation can lead to an extreme wildfire risk across regions not accustomed to such aridity. Awful spells of high heat and dry conditions have led to devastating wildfires across the Australian outback in recent years. California has been particularly hard-hit by drought-induced infernos; the state witnessed 5 of its 10 largest wildfires on record during the summer of 2020, with 2021 on trend to outpace it.

cold snap to descend over Orlando, Florida, even though the popular tourist destination is usually warm and humid for most of the year.

It's tougher to issue long-term climate predictions than it is to accurately predict the weather several days out. There are so many intricate and fragile processes that go into long-term patterns that it's nearly impossible to say months in advance that it'll be precisely 83°F on Memorial Day. However, weather models are powerful enough today to suss out patterns that make it possible for long-range forecasters to spot that temperatures could be above average for the end of May.

The science of long-range climate predictions is wonderfully useful for every slice of life. We encounter these forecasts during the fall months, for instance, when meteorologists at local television news stations issue their annual winter weather outlooks. The same is true ahead of the summer months, when experts release their hurricane outlooks, letting coastal residents know if an upcoming hurricane season could be exceptionally active or blissfully quiet.

While long-range forecasts are useful for a heads-up about a snowy period or a raucous bout of hurricanes, climate records and long-term climate forecasts are also

## AN IMPENDING CATASTROPHE

Climate change is more than just an inconvenience marked by occasional storms. Long-term pattern changes will have tangible and devastating effects around the world. Longer droughts could stress water resources to the limit in the American Southwest, where growing communities are already burdening the region's rivers and lakes. Coral reefs are dying as a result of warmer ocean temperatures, resulting in significant and ongoing damage to Australia's Great Barrier Reef. Rising ocean levels, heavier rain events, and more rapid intensification of hurricanes gravely threaten low-lying cities such as New Orleans, which could experience intense freshwater and seawater flooding as a powerful storm pushes inland.

flashing warning lights for what lies ahead: our climate is changing. Summers are growing hotter and winters are getting shorter. Intense rain is falling more heavily than ever before, and dry spells are spiraling into destructive droughts.

Earth's climate has always changed. We can see the evidence of this in the fact that the vast majority of Canadians now reside comfortably where miles-thick glaciers once covered the land. Indeed, the entire United Kingdom, most of Canada, and a large part of the northern United States were blanketed by glaciers as recently as 18,000 years ago. These glaciers carved and filled the Great Lakes. Today, as a result of long-term climate change, these ice sheets have retreated almost entirely to the Arctic and Antarctic Circles, leaving behind nourished land beneath a temperate climate.

It used to take many thousands of years for Earth's climate to cycle between warm spells and cold spells. Those days of slow climate change are long gone. Alterations in long-term weather patterns that used to take hundreds or thousands of years are now perceptible in our lifetimes. We notice it in ways both small and large. Older relatives often remark that summers are hotter today and winters aren't quite so cold as they used to be. Hurricanes stall more frequently and produce heavier rain than they once did.

Today, the climate is changing at an alarming rate, and the overwhelming majority of scientists agree that the change is anthropogenic, or directly influenced by human activities. Centuries of pollution have gushed into the atmosphere from enterprises like motorized transportation, coal-fired power plants, and the slashing and burning of untold acres of land.

The untenable excess of pollutants in the atmosphere has led to a phenomenon known as the greenhouse effect. Named after the glass-enclosed structures that provide a warm environment for plants to grow, the greenhouse effect describes how warming pollutants—known as greenhouse gases—efficiently trap solar radiation in the atmosphere, slowly forcing the planet to warm.

Earth's atmosphere is currently packed with more greenhouse gases than at any time in the course of human history. The most prolific greenhouse gas we emit in our atmosphere is carbon dioxide. Scientists can study samples such as ice cores to measure the amount of carbon dioxide in Earth's atmosphere dating as far back as 800,000 years ago. During that period, the amount of carbon dioxide in the atmosphere never exceeded 300 parts per million. The level of carbon dioxide in the atmosphere soared during the twentieth and twenty-first centuries, spiking to nearly 420 parts per million

# CLIMATE CHANGE

FOREST FIRES

RISING SEA LEVELS

BLEACHING CORAL

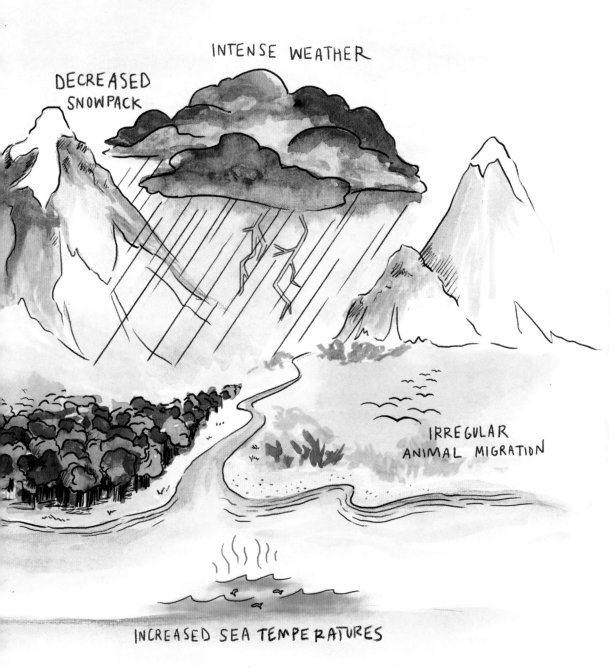

DECREASED SNOWPACK

INTENSE WEATHER

IRREGULAR ANIMAL MIGRATION

INCREASED SEA TEMPERATURES

Climate change will have widespread and deeply interconnected effects around the world, ranging from heat waves and droughts that spark intense wildfires to rising sea levels that submerge bustling cities.

by 2020. Climatologists have long warned that carbon dioxide levels could reach a point of no return, inducing rapid climate change that would provoke significant weather disasters in the future.

The effects of climate change don't seem like much at first glance. Average annual temperatures around the world rose about 2°F between the end of the 1880s and the end of the 2010s. That might not appear to be much of a jump in average temperatures, but climate change is akin to a fever in the human body. While Earth's climate is obviously much more complex than the human body, consider how relatively little your body temperature has to rise for you to go from feeling fine to deliriously feverish.

Extensive warming is already an issue in polar climates, where temperatures have risen dramatically in recent decades. Sea ice in the Arctic Circle has rapidly declined every decade since 1980. The decline was so precipitous that, by 2018, a cargo ship was able to sail from South Korea to Germany through the Arctic Circle.

Not only will melting sea ice result in even warmer temperatures in the Arctic, which could destabilize global weather patterns even further, but warmer air temperatures in polar regions will induce the ice sheets over land in Greenland and Antarctica to melt. Sea levels don't rise when sea ice melts—for the same reason a glass of ice water doesn't overflow as ice cubes melt—but runoff from melting ice on land will raise sea levels to disastrous effect. Rising ocean levels will flood coastal communities around the world. Coastal flooding is commonplace now during high tide in communities along the East Coast of the United States, including Miami, Florida, and Charleston, South Carolina.

A warmer atmosphere with more frequent extreme weather events will force some areas into extensive droughts while other areas will struggle to deal with flooding rains. We already see evidence of extreme flooding from hurricanes that hit the United States. Hurricane Matthew in 2016, Hurricane Harvey in 2017, and Hurricane Florence in 2018 each broke all-time rainfall records for tropical cyclones when they made landfall in the southeastern United States.

Mitigating the effects of climate change and preparing for the dangerous extremes it has already unleashed is a herculean challenge that countries around the world are scrambling to meet. It's still possible to stem some of the worst-case scenarios that could result from rapid and unchecked climate change, but it'll take intense focus and serious resolve from both politicians and populations to accomplish.

# DISAPPEARING ISLAND NATIONS

The water rising from climate change will pose a grave threat to some island nations, including the low-lying islands of the Maldives, where most of the population lives within just a few feet of sea level. The entire country could disappear by 2100.

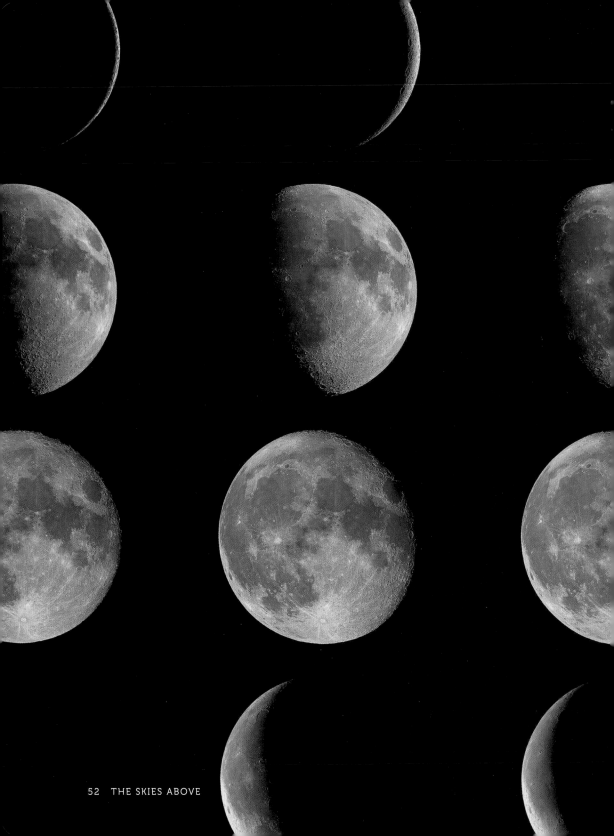

# THE LUNAR PHASES

**F**our and a half billion years ago, a huge object hurtling through space collided with Earth. Bang! The impact shattered both our young planet and the rogue body that crossed its path, ejecting debris from each out into space. While this scenario could have been catastrophic (just ask the dinosaurs), this earthly collision led to something lovely. Over millions of years, some of the debris still orbiting Earth coalesced into a satellite we've come to know as the Moon.

Earth isn't alone in its companionship with a moon. While we simply call Earth's moon "the Moon," any large natural satellite orbiting any planet can be considered a moon. The largest moon in the solar system is Jupiter's Ganymede, which is nearly 50 percent larger than our own. Many of the moons are tiny, such as Mars's Phobos, which is so diminutive that it could comfortably fit within the shores of Lake Erie with room to spare.

Hundreds of moons orbit most of the planets in our solar system—except Mercury and Venus. We can see some other planets' moons in the night sky if we look hard enough. On a clear, dark night, a basic telescope, a good pair of binoculars, or even the zoom feature on a steadied point-and-shoot camera can reveal moons orbiting Mars, Jupiter, and Saturn.

But none is more spectacular in the night sky than our own. The proximity and prominence of the Moon in our sky has guided the course of human civilization since it began, serving as the muse and basis for religions, artistic masterpieces, and the calendars that have guided our perception of time. But, more than that, the Moon's place as the Earth's trusty sidekick helps make life on our planet possible.

Life would go a little sideways without the Moon. The gravitational pull from our natural satellite helps stabilize the Earth's axial tilt, ensuring that we won't start toppling end over end—at least, not in our lifetimes. The pull of the Moon's gravity also induces tides on the world's bodies of water. Tidal systems are essential to marine life, coastal environments, and communities that rely on the ocean for their lives and livelihoods. A close tracking of tides is also important in weather forecasting, since coastal flooding from a landfalling hurricane can be even worse if the storm makes landfall at high tide.

If we were to look at the Earth and the Moon from above, the Moon would appear to orbit counterclockwise around the Earth. The Moon rotates at the same rate as it revolves around the Earth, so we always see the same side of the Moon. A slight wobble in the Moon's axis as it orbits, combined with where we are on the Earth's surface, changes how the Moon looks in the sky—but it's always the same side, no matter how we look at it.

The slow orbit of the Moon around the Earth means that the Moon's illumination appears a little different in the sky each day. We mark these differences at eight different points, or phases, of the Moon's orbit.

All told, one orbit of the Moon takes a little over 29 days to complete. The 12-month calendar we use today, known as the Julian calendar, doesn't follow the lunar cycle

# PHASES OF THE MOON

Following is a description of the phases of a sample month.

| DATE | PHASE | PERCENT ILLUMINATED | |
|------|-------|---------------------|---|
| January 1 | Full Moon | 100% | A full moon occurs when the Earth is situated between the Moon and the Sun, illuminating the lunar surface in the Earth's night sky. |
| January 5 | Waning Gibbous | 75% | |
| January 8 | Third Quarter | 50% | The third quarter, the aptly named half moon, appears half illuminated and shows up for about six hours during the day and six hours at night. |
| January 11 | Waning Crescent | 25% | |
| January 16 | New Moon | 0% | A new moon occurs when the Moon is situated between the Earth and the Sun, illuminating the side of the Moon we can't see from the Earth. |
| January 21 | Waxing Crescent | 25% | |
| January 24 | First Quarter | 50% | The first quarter is the second half moon of the month, with its illumination appearing as the mirror image of the third quarter because it's on the other side of the Earth. |
| January 26 | Waning Gibbous | 75% | |
| January 31 | Full Moon | 100% | Every two or three years a second full moon—called a blue moon—may occur in a month. |

# PHASES OF THE EARTH

When the astronauts of Apollo 8 circled the Moon in 1968, they made history as the first humans to leave low-earth orbit. Looking at Earth from afar was a perspective no human had ever witnessed before. Our planet, perched against light years of darkness, looked so lonely and so fragile that a snapshot of their view of a partially illuminated Earth over the lunar landscape, "Earthrise," soon grew into one of the most famous images ever taken.

The phases of the Earth from the lunar surface are the exact opposite of the phases of the Moon on the Earth. When we see a bright full moon on the Earth, folks on the Moon would see a "new Earth," gazing up at the gentle glow of nighttime on our home planet. Likewise, the day of a new moon on the Earth would occur during nighttime on the Moon, giving lunar-dwellers a grand view of the full Earth in the night sky.

The day in 1968 when those intrepid astronauts orbited the Moon for the first time, a waxing crescent moon could be viewed from the Earth, so only about a quarter of the Moon was visible for a small portion of the night to folks back home. In "Earthrise," however, our planet is about three-quarters illuminated.

like many previous calendars did. Since each month is longer than the time it takes for the Moon to complete a single revolution, the misalignment between our temporal accounting scheme and the Moon's movement results in lunar phases occurring on a different date from one month to the next. Sometimes, we can see the same phase twice in a month.

## Unusual Full Moons

If we're treated to a full moon twice in one month, it's called a blue moon. Seen only once every couple of years, its appearance gives rise to the phrase "once in a blue moon" to describe something that rarely occurs. While the term likely originated in sixteenth-century literature as an exaggerated mockery—sort of how folks today say "two plus two equals five" to deride someone who's obviously lying—it's not clear when the term came to describe some-thing rare or improbable.

# THE MOON'S NOT-SO-GREAT ESCAPE

The Earth and the Moon depend on each other for their survival. The Earth needs the Moon to keep a stable axis and remain habitable for life, while the Moon needs the Earth's gravitational pull to keep it from being flung into space. Thankfully, we don't ever have to worry about that orbital bond breaking in our lifetime (or the many dozens after that). The two bodies are in such a stable relationship with each other that the Moon is creeping away from the Earth at a rate of only about one foot every decade. It will take millions of years for this trend to appreciably alter the lunar orbit.

A blue moon isn't the only full moon with a special name. There's also the much-vaunted supermoon, which is a larger-than-normal full moon in the night sky. This is the result of a glitch in the Moon's orbit, which isn't perfectly centered on the Earth. Its closest approach to our planet, called the perigee, places the Moon at about 220,000 miles from the Earth's surface, which results in this special full moon. NASA says the full moon looks only about 18 percent bigger at perigee, a difference that's nearly imperceptible to the naked eye, but a supermoon appears larger in photographs when compared to photos of other full moons. A full moon that occurs during farthest outreach of the lunar orbit, called the apogee, is sometimes called the micromoon. This full moon will appear only about one-tenth smaller in the night sky.

Full moons are culturally significant around the world. The origin of the names they're given often derives from cultures that relied on the phases of the Moon to keep track of time, so the names are closely related to activities or observations made around that particular instance of the full moon. A harvest moon, for instance, usually coincides with the fall harvest in September or October. A strawberry moon makes its appearance in June, and the cold moon rises on a frigid December's night.

> The moon is like a scimitar,
> A little silver scimitar,
> A-drifting down the sky.
> And near beside it is a star,
> A timid twinkling golden star,
> That watches likes an eye.
>
> —SARA TEASDALE, "Dusk in Autumn"

## THE DAYTIME MOON

The reason the Moon is still visible during the day is that the surface of the Moon isn't completely dark. Just as the full moon hangs in our night sky like a spotlight, sunlight reflecting off the Earth's brilliant surface casts a blue shine on the Moon's surface during a new moon.

# LUNAR ECLIPSES

A breathtaking astronomical event, a lunar eclipse is one of the easiest to enjoy. You don't need a telescope, a pair of special glasses, or complete darkness in order to witness the surface of the Moon slowly fading to a dark shade of red. As long as you have a clear view of the sky without any clouds in your way, you can easily see the Earth's shadow obscure the Moon.

A lunar eclipse is the result of a syzygy, a delightfully fun word that describes the linear alignment of the Sun, the Earth, and the Moon. This occurs when the Earth moves between the Sun and the Moon, casting its shadow on the lunar surface. The size of the Earth relative to the Moon means a lunar eclipse lasts a couple of hours from beginning to end, giving observers plenty of time to head outside and enjoy the view.

Atmospheric filtering
of shortwave light
(cool blues)

SUNLIGHT    PENUMBRA    UMBRA

EARTH    MOON

Longwave light visible
(warm oranges & reds)

We see a lunar eclipse when the Moon passes through Earth's umbra, or the darkest part of the planet's shadow. The Moon appears red at totality because of light filtering through our planet's atmosphere.

Earth's shadow isn't uniformly dark. The Sun is enormous, so our planet receives sunlight from every visible point on the star's surface. In addition to direct sunlight, we also receive light on a slight angle from the outer edges of the Sun's disk. This causes the Earth to cast a shadow on the Moon that's faint on its edges and grows progressively darker toward the middle. You can see this effect in your own home. When you turn on a light, objects near it don't cast a single, sharp shadow on the wall; instead, they cast a dark shadow that gradually grows fuzzier and fainter around the edges.

To get technical, the faint edges of the Earth's shadow are called its penumbra, while the dark center of the shadow caused by blockage of the Sun's direct light is called the umbra. A total lunar eclipse takes place when the Moon passes directly through the umbra, while a partial lunar eclipse sees the Earth's umbra clip only a portion of the lunar surface. Penumbral eclipses are more common than umbral eclipses, but the shadow is so faint that it's hard for most people to notice a difference in the brightness of the Moon.

Total lunar eclipses, or umbral eclipses, are the most impressive of the bunch. The first glimpse of our planetary obstruction is a thin crescent shadow that appears on the corner of the Moon. This sliver of darkness will eventually grow to encompass the

# BLOOD MOON

A total lunar eclipse is occasionally called a blood moon due to the reddish tint that bathes the Moon as it falls behind the Earth's shadow. The term grew popular after Christian evangelical pastor John Hagee released the book *Four Blood Moons: Something Is about to Change* in 2013. The pastor argued that the four total lunar eclipses visible in 2014 and 2015 were fulfillment of a Biblical prophecy that heralded the impending arrival of the end times. While the term's modern popularity apparently began after the release of Hagee's book, it's common for news organizations to refer to lunar eclipses as blood moons as a snappy way to advertise the astronomical event on social media.

entire body itself, bathing the lunar surface with a dark red shadow that looks as though the Moon were on fire. In some cases, totality can last more than an hour before the shadow moves on, allowing the Moon to slowly regain sunlight until its orbit fully escapes the Earth's umbra.

The reddish tint on the Moon during the height of a total lunar eclipse occurs for the same reason we see a beautiful display of colors at sunrise and sunset. Sunlight scatters as it filters through Earth's atmosphere. The individual waves of visible light emanating from the Sun bump up against all the gases and pollutants that fill our atmosphere, slowly filtering out shortwave light like the color blue, while longwave colors like orange and red are able to slip through without getting filtered out. Sunlight during an eclipse has to pass through almost the entire atmosphere to reach the Moon, allowing only the deepest shades of red to reach the lunar surface.

Lunar eclipses are possible only during a full moon, or when the Moon is on the dark side of the Earth. The full moon appears in the sky at sunset and sinks below the horizon at sunrise. The bright moonlight that bathes the Earth's surface with its revealing glow doesn't come from the Moon itself but rather from sunlight reflecting off the dust and rocks that make up the lunar surface. It's impressive how brightly the Moon can illuminate the night sky given that it's only a reflection!

If we can see a lunar eclipse during a full moon, then why don't we see them every month? The Moon's orbit around the Earth isn't flat; it's actually tilted by a couple of degrees, meaning the Moon passes above or below the Earth's shadow most of the time. The Moon's orbit rarely takes it through some or all of our planet's shadow.

The Moon's small size and slightly tilted orbit makes solar eclipses even rarer than lunar eclipses. A solar eclipse occurs during the new moon, when the Moon passes directly between the Sun and the Earth at just the right angle so that the Moon casts a shadow on the Earth's surface.

As each total lunar eclipse is visible by about half of the world, the average person will likely have the opportunity to see dozens of lunar eclipses during their lifetime. Seeing a total solar eclipse is much rarer because the Moon's shadow covers only a relatively tiny portion of the Earth's surface. On average, any one spot on Earth gets to see a total solar eclipse once every 375 years, but the same spot has the chance to witness a total lunar eclipse once every two years or so.

# TOTAL SOLAR ECLIPSE

Anyone near the path of a solar eclipse can use protective glasses to watch the Moon slowly cover the Sun, making our star appear as a vanishing crescent in the sky. The Sun is so bright that even 80 to 90 percent coverage still provides as much sunlight as you'd see on a cloudy day. The moment when the Moon completely covers the Sun in the sky is called totality. Locations covered by totality experience nighttime-level darkness for up to a couple of minutes. Observers can even see the hazy glow of the Sun's atmosphere emerge from behind the edges of the Moon.

# ASTEROIDS, COMETS, AND METEORS

**A** deep gaze into the night sky reveals countless stars that twinkle and swirl as Earth slowly spins through the vast realm of space. We're certainly not alone out here, and not everything in our sky is as far away as it appears. The deep darkness of a still night can break in a bright streak that ends in a brilliant flash illuminating the landscape like daylight. Each generation is treated to at least one brilliant display of a comet's feathery tail as the icy body whirls around the Sun. And a handful of unlucky humans have witnessed the dark side of these cosmic collisions, such as the enormous space rock that damaged thousands of buildings when it exploded in a fireball over a Russian city in 2013.

While we know our solar system as the home to eight planets and hundreds of moons, there are millions of smaller objects orbiting around the Sun as well. These smaller objects—including asteroids, comets, and meteoroids—all play an interesting role in astronomy, and they interact with Earth in vivid and sometimes dramatic ways.

## Asteroids

Asteroids are large rocks floating through space that can contain silicate rocks, clays, metals, and frozen liquids and gases. They're smaller than planets or moons, but some are large enough for astronomers to name them and plot out their orbits. These jagged bodies are the byproduct of eons of galactic billiards: planets colliding with planets, comets hitting planets, moons crushing one another. The ejecta produced by each of those huge collisions resulted in most of the asteroids that meander the solar system today.

The size and shape of an asteroid can vary from the size of a school bus to about one-quarter the size of Earth's moon, which is the size of Ceres, the largest asteroid in the solar system. Ceres, like Pluto, is classified as a dwarf planet—an object orbiting the Sun that's neither a full-size planet nor a moon, but is large enough to generate the gravity necessary for the body to take on a rounded shape.

## WHY DINOSAURS HATE ASTRONOMY

One of the largest confirmed impact craters on Earth likely resulted in a very consequential event for life on the planet. The Chicxulub crater is a scar on Earth's surface that measures nearly 100 miles across. The meteorite—a large asteroid or comet that survived the atmosphere intact—hit in the southern Gulf of Mexico near the northern tip of Mexico's Yucatán Peninsula. The immense shockwave, fallout, and ensuing climate disruption from this impact contribute to a leading theory about the cause of the mass extinction of the dinosaurs.

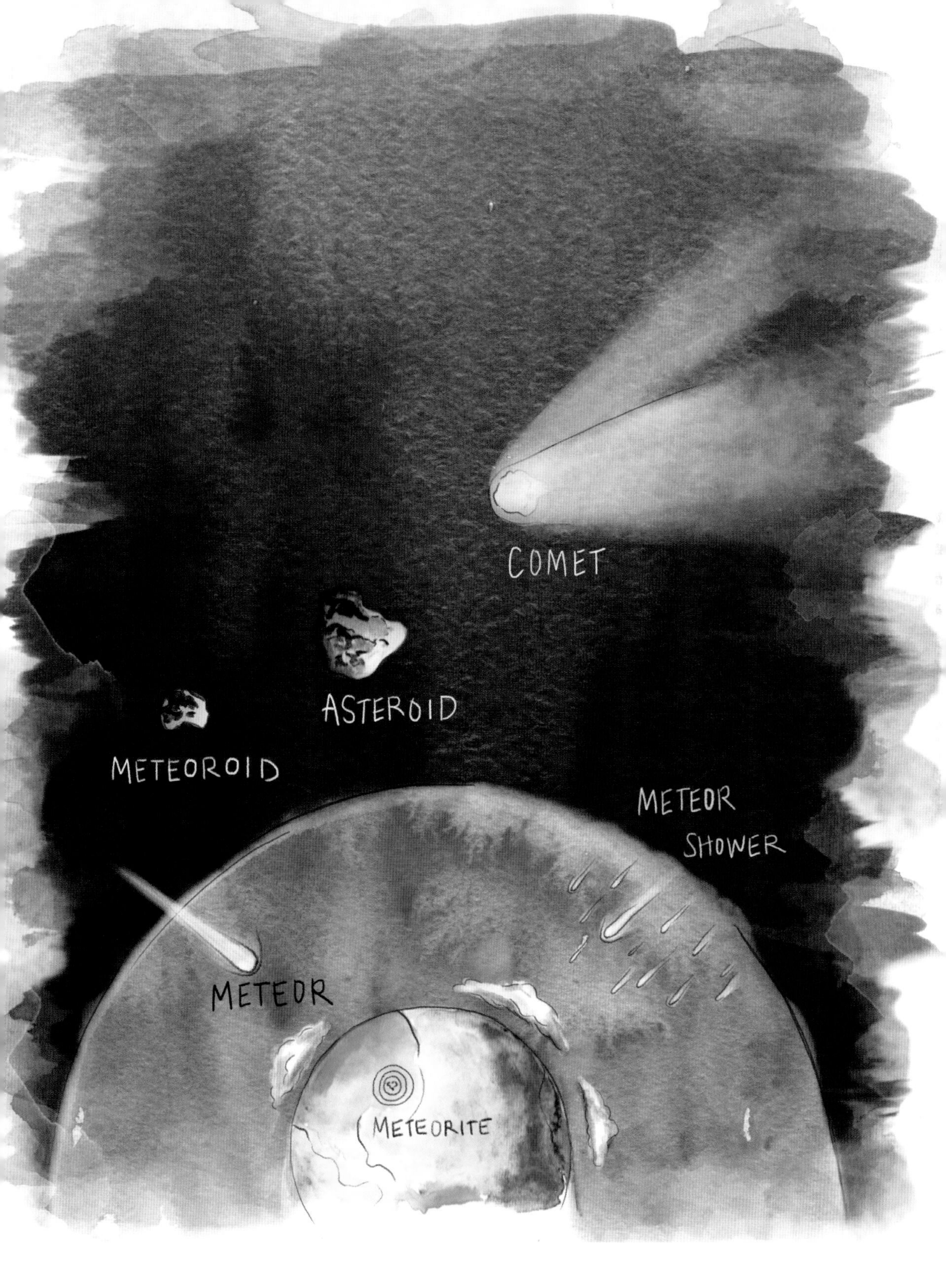

COMET

ASTEROID

METEOROID

METEOR SHOWER

METEOR

METEORITE

Over time, millions of asteroids have fallen into a somewhat stable orbit around the Sun, coming to occupy the large gap between the orbits of Mars and Jupiter. This region, called the asteroid belt, contains the vast majority of known asteroids in our solar system, including Ceres. One theory holds that these asteroids would have formed a new planet had it not been for a combination of asteroids shattering when they hit one another and the gravitational pull of Jupiter keeping the larger asteroids from coalescing into a newborn planet.

Most asteroids are contained within the asteroid belt, but some are loose in the solar system and plot their own course. Every once in a while, a collision destabilizes an asteroid's orbit and flings it outside of the asteroid belt. Many of these wayward asteroids are tracked by astronomers as potential hazards to the Earth or the Moon. NASA's Center for Near Earth Object Studies (CNEOS) is an effort to keep tabs on asteroids that could potentially crash into Earth. The agency recorded nearly 24,000 near-Earth asteroids through the end of 2020, with nearly 9,500 of these close objects measuring at least 500 feet across.

## Comets

Comets provide some of the most memorable sky-watching experiences for folks lucky enough to catch a glimpse of these slow-moving displays in the night sky. The largest comets can grow to the size of an entire town, measuring more than a dozen miles across. These objects are composed of rocks, frozen gases, and frozen liquids. Scientists often compare comets to snowballs made from street slush because of their dirty composition. Much like an asteroid, a comet can inflict serious damage on moons or planets if they cross paths with one of these icy behemoths.

Most comets are unremarkable when they're deep in the solar system. It's when a comet's orbit brings the object close to the Sun that things get interesting. When we look at a comet from Earth, we're not seeing the comet itself, which is much too small to see with even a good telescope. The visible parts of a comet are its gas tail and dust tail, both of which are the result of the Sun's warmth.

Direct sunlight heats up the surface of a comet as it approaches the Sun, causing the frozen gases within a comet to sublimate—transition directly from a solid to a gas—and vent out into space. The gas tail flows from the comet in a straight line away from the Sun, while the dust tail—comprised of all the rocks and dust particles released or

I forget not to sing;
Nor the comet that came unannounced,
    out of the north,
flaring in heaven,
Nor the strange huge meteor procession,
    dazzling and
clear, shooting over our heads,
(A moment, a moment long,
    it sail'd its balls of unearthly light over our heads,
Then departed, dropt in the night,
    and was gone;)

—WALT WHITMAN, "A Year of Meteors"

knocked loose when the gas sublimates—exits the comet on an angle, leaving a vast, feathery tail in the comet's wake.

The path a comet takes as it approaches the Sun affects its visibility on the Earth. Some comets orbit the Sun on a regular basis, providing a predictable display in the night sky over a set period of time, while others appear only once and are never seen again.

The orbit of Halley's Comet, for instance, brings the body into Earth's view once every 75 years or so; it last came around in 1986, and it'll reappear in the night sky in 2061. Comet Hale-Bopp, on the other hand, tracks along such an elongated orbit that the spectacular display it provided in 1997 and 1998 won't happen again until well after the year 4000.

# SPACE DUST

If you've ever taken in the sight of a starry sky and noticed a strange, bright patch of fuzziness adorning the vast expanse of space, you've probably witnessed space dust. Also known as interplanetary dust, this collection of tiny particles is shed by asteroids and comets as they've collided and broken apart over the eons. The dust earns its faint glow from sunlight scattering through the particles. Scientists call this phenomenon the zodiacal light because it appears in the sky at roughly the same angle as the Sun and other planets. The glow of space dust in the night sky is so faint that it can be detected only in complete darkness.

# Meteoroids

Those asteroids and comets leave quite a bit of cosmic litter in their wake. Each asteroid that breaks apart jettisons countless little bits of rock into space. Each of those chunks of rock can hit other asteroids, chipping them and creating more debris that can continue the chain reaction down the line. All the dust sprayed into space by the tail of a comet is packed with small fragments of rocks, metals, and particles that are free to move around and strike what they please.

Those tiny fragments of rocks and metals are meteoroids. Meteoroids can be as small as a grain of sand, or they can measure many hundreds of feet across. It's nearly impossible to keep track of most meteoroids while they're still in space. This is a harrowing prospect for humankind's ever-growing network of satellites and spacecraft, which can be seriously damaged or even destroyed by meteoroids the size of a pebble. Scientists attempt to track larger meteoroids to keep astronauts safe; these efforts have given folks aboard the International Space Station enough time to get out of the way of wayward meteoroids and other space debris that could threaten their safety.

# Meteors

Any meteoroid or small asteroid that falls through Earth's atmosphere is a meteor. Commonly called shooting or falling stars, a chance encounter with a meteor high in the dark sky feels like a magical experience. After all, watching a piece of space rock zip through the atmosphere is the closest most of us will ever get to touching space without hopping the next rocket into orbit.

Meteors move fast. When these speedy objects hit Earth's atmosphere, the intense friction along the leading edge of the meteor turns the air into plasma, much like the fiery glow that appears outside of space capsules on reentry. The blazing heat of entry into the atmosphere causes the meteor to burn and break apart. It takes only a couple of seconds for a meteor to hit the atmosphere, burn up, and disintegrate.

What results is just spectacular. Meteors can appear different colors, depending on the speed of the meteor when it hits the atmosphere and what metals are contained within the rock. Meteors packed with magnesium emanate a bluish light as they burn, and sodium produces yellow hues. (Burning metals producing different colors is the same principle that pyrotechnicians take advantage of to create fireworks.)

## Meteor Showers

When a comet passes through or near Earth's orbit, the dust tail left behind by the comet is a reliable source for meteors. These predictable bouts of meteors are meteor showers. The quality and quantity of a meteor shower depends on the thickness and density of the dust tail through which Earth passes.

Meteor showers are usually named after the constellation—a group of stars that forms a recognizable pattern in the sky—from which the meteors appear to originate in the night sky. The Perseids are a midsummer meteor shower produced by the dust tail left behind by Comet Swift-Tuttle. Meteors during the Perseids appear to enter the atmosphere near the Perseus constellation in the Northern Hemisphere's north sky.

The best viewing for a particular

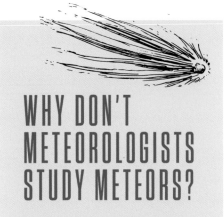

# WHY DON'T METEOROLOGISTS STUDY METEORS?

The word "meteor" itself is one of life's confusing little pleasures for weather geeks. Astronomers study meteors, but meteorologists study the weather. Meteors also refer to atmospheric phenomena such as rain, snow, and hail. Meteor derives from a Greek word that means "high in the air," a fitting—and possibly understated—loanword for falling bits of space rock. Meteorologists may find fascination in space meteors, but the association is merely linguistic.

meteor shower depends on the orientation of Earth's axis when it passes through a comet's dust tail. The hemisphere that leads headfirst into the dust cloud, so to speak, sees the best display of meteors. The Leonids and Perseids are concentrated in the Northern Hemisphere, while the Delta Aquariids are best viewed in the Southern Hemisphere, where the Aquarius constellation appears high in the sky in July and August.

The Leonids are a fantastic meteor shower that everyone should have the pleasure of watching at least once. This mid-autumn event occurs when Earth spins through the dust tail of Comet Tempel-Tuttle. The Leonids, appearing near the Leo constellation, can produce more than 100 meteors per hour; such a prolific bout of meteors is often called a meteor storm. Scientists can estimate when meteor storms are possible based on the density of the dust through which Earth will pass. Sometimes these

# IMPACT CRATERS

Larger meteorites will leave well-defined impact craters on Earth's surface that can survive weathering for tens of thousands of years. The Arizona desert is home to one of the best-preserved meteorite impact craters on the planet. The aptly named Meteor Crater was created approximately 50,000 years ago, is nearly three-quarters of a mile wide, and sinks more than 100 feet into the desert floor.

predictions are wrong, while some meteor showers can produce many more meteors than forecast and catch viewers pleasantly by surprise.

## Fireballs

Some meteors are bright, and others are dim; some leave behind tails that stretch across the whole sky, while others seem to produce no tail at all. But no matter how they enter or what they leave behind, some of those run-of-the-mill meteors end in a bang. Occasionally, a meteor will explode in a vivid fireball that can light up the sky with the ferocity of a second sun.

A fireball, formally called a bolide, is mostly harmless, creating only a bright flash in the sky that can startle drivers and catch the attention of folks who happen to be outside at the right moment. Some fireballs, though, aren't quite so innocent.

In 2013, the residents of Chelyabinsk, Russia, found themselves under the path of a meteor that measured more than 60 feet across. This meteor crashed through the sky over southwestern Russia during the morning commute, creating a flash brighter than the Sun as it plowed into the atmosphere. The fireball, which left a long, lingering condensation trail in its wake, generated a sonic boom when it exploded. The air blast from the explosion damaged thousands of homes and businesses across Chelyabinsk and neighboring communities, injuring hundreds of people from the blast and the flying debris it created.

## Meteorites

Pieces of the Chelyabinsk meteor made it down to the ground. Meteors that survive their tangle with friction and make it to Earth's surface are called meteorites. Most meteorites are tiny enough to pick up with one hand, since meteors break apart or explode during the intense heat of atmospheric entry. Larger meteors can survive their encounter with the atmosphere and hit the ground with a grandiose thud.

Museums around the world are packed with samples of meteorites that were found lying on the surface or dug up after thousands of years buried by soil. A farmer in central Kansas found a meteorite buried on his property that weighed nearly 1,000 pounds. The rock went on display at a museum in Greensburg, Kansas, where it survived another one of nature's intense forces: the scale-topping tornado that destroyed the town in 2007. Officials recovered the meteorite from the debris and returned it to the rebuilt museum.

Bears have the right idea to sleep through the winter. Long nights and bitterly cold days make winter an unpleasant experience for folks who live at higher latitudes. Winter is all about the snow and cold—blizzards, nor'easters, crusts of ice, frigid winds, and the much-maligned polar vortex. Yet, the sky above still provides some beauty to get us through the long slog of an ugly winter. Icy clouds catch and bend the fleeting sunlight into brilliant arcs and rainbows high in the sky, while solar radiation seeping through Earth's magnetic field puts on a vivid light show against the perpetual night sky in the north.

# THE WINTER SOLSTICE

**W**inter is both bleak and beautiful. Short days and frosty temperatures act like a natural reset for life on our planet. Plants shed their leaves to prepare for the long wait for warmer days that lie ahead. We ourselves tend to hibernate, slowing down and finding new hobbies to occupy our time while we weather cold spells and driving snow. But winter isn't all hiding and waiting for the Sun to shine again. There's plenty to admire even when conditions are harsh. Gorgeous auroras light up the night sky. Shimmering halos paint the icy clouds that hang high on a chilly day. White blankets of snow and coats of ice glisten on the ground and highlight the beauty of the land around us. It's easy to find wonder in winter, and we owe it all to the good fortune of Earth winding up a little lopsided in its orbit.

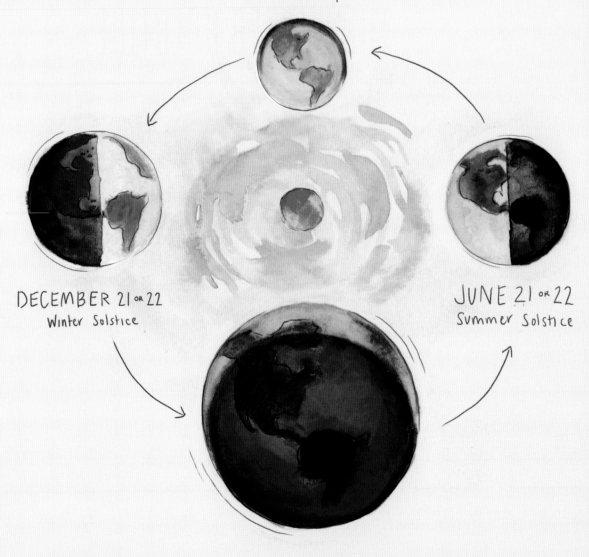

AROUND
SEPTEMBER 22
Autumnal Equinox

DECEMBER 21 or 22
Winter Solstice

JUNE 21 or 22
Summer Solstice

AROUND MARCH 20
Vernal Equinox

## Earth's Tilting Axis

The Earth receives light and warmth from all visible points of the Sun, from top to bottom and from side to side. The most intense radiation reaches the Earth from around the Sun's equator. After all, the Sun is enormous—it's more than 100 times bigger than the Earth itself—so the geographic center of this star is also its closest point to the Earth.

However, our planet's equator doesn't line up with the Sun's equator. The uneven heating of Earth's surface that results from the planet's axial tilt allows four distinct seasons: winter, spring, summer, and fall. If our planet didn't rotate on an angle relative to its orbit, things would look much different around here. The North and South Poles, would never experience meaningful sunlight, and the greatest solar energy would always point directly at the equator.

Humans found ways to adapt to extreme heat and extreme cold long before the age of electric heaters and air conditioning. But a planet without seasons, an axial tilt that's straight up and down instead of agreeably askew, would be unbearable. The monotony of direct sunlight every day would make tropical areas and deserts nearly uninhabitable from excessive heat, and polar regions would wind up similarly unlivable due to the relentless darkness and harsh frigidity.

Earth's rotation takes care of that problem for us. The planet spins along an axis that tilts about 23.5° relative to its orbit, so the combination of spinning every 24 hours and rotating around the Sun once every 365.25 days

> What good is the warm of summer, without the cold of winter to give it sweetness?
>
> —JOHN STEINBECK, *Travels with Charley*

(hence the leap day every four years) means that the equator itself receives direct solar radiation only twice a year. The Earth's steady, speedy orbit ensures that the Sun's rays aim for a different spot each day.

The angle between the Sun's equator and the Earth's equator is the solar declination. Direct sunlight shines down on 23.5°N on the late June day when the North Pole points toward the Sun and the solar declination reaches its northernmost extent. When the Sun's energy focuses on this latitude, called the Tropic of Cancer, it's the summer solstice in the Northern Hemisphere and the winter solstice in the Southern Hemisphere.

Half a year and two seasons later, the southernmost extent of solar declination occurs in late December, when sunlight shines strongest on 23.5°S, a line nicknamed the Tropic of Capricorn. These names came about in ancient times when the Sun appeared in the same portion of the sky as the eponymous constellations during the two solstices.

If you stood on either of those two special latitudes during their respective summer solstices, the Sun would appear directly overhead, with tall objects casting no shadow on the ground. If you were to stand outside in the same spot on the winter solstice, you would cast the longest shadow of the year as the Sun hangs low in the sky.

## Months of Darkness

Long shadows are the very least of the darkness cast by the winter solstice. The colder months are rough for folks who thrive on warmth and sunshine. Days leading up to the winter solstice grow noticeably shorter as the Sun sinks lower on the horizon. The length of the day depends entirely on latitude. Boston, Massachusetts, which is located

# THE DANGERS OF COLD

Cold air means more than just chapped lips and dry hands. Extremely cold temperatures can seriously injure the human body in short order. Frostbite sets in when human flesh freezes, leading to injury as mild as a persistent burning sensation to a medical emergency that requires surgical intervention. Prolonged exposure to extreme cold can also lead to hypothermia, a serious condition that occurs when the body's temperature drops dangerously low.

It takes only 30 minutes for frostbite to develop on exposed skin when the air temperature drops below about -15°F. Temperatures of about -30°F can lead to frostbite in just 10 minutes, and polar conditions where temperatures fall below -50°F can cause frostbite in as little as 5 minutes. These temperatures are rare in, say, New Mexico, but they're not hard to come by during the heart of a winter cold snap in northern latitudes.

## STEAM FOG

Folks walking near a pond or a lake on a brisk morning can witness a mesmerizing display of fog rising from the water and dancing above the surface. These wisps of condensation are the result of warm, moist air in contact with the water rising and condensing through the colder air above. This miniature form of convection results in a thin layer of fog that looks like steam.

# WIND CHILL

A gust of wind can cut right through you when it's cold out. Wind really does make the air feel colder than it is. Meteorologists measure this phenomenon using the wind chill index. When you step outside on a frigid day with calm winds, you're somewhat insulated from the elements because your radiant body heat builds up a thin layer of warm air right above the exposed skin. This helps keep you warm for a little while before the chill takes its toll. Gusty winds force that cold air to make direct contact with your skin, chilling your exposed areas much faster than they would cool down in calm air.

Factoring in wind, a 0°F morning with a 15-mile-per-hour wind can have the same effect on your body as an air temperature of -19°F. Even though the actual temperature is significantly warmer than -19°F, the wind buffeting against your exposed skin can lead to injury in 30 minutes compared to the hour or longer it would take for the cold to hurt your face or hands on a calm single-digit morning.

around 42.4°N latitude, receives about nine hours and four minutes of sunlight on the winter solstice. Bostonians spend a solid chunk of winter commuting to work in the dark and heading back home when it's dark. It's a rosier story down south. Miami, Florida, which lies about 1,500 miles south of Boston and very close to the Tropic of Cancer at 25.8°N, experiences a full hour and a half more sunlight on the winter solstice.

Sunlight shifting and shortening the days as winter settles in doesn't just make things rough on folks in the middle latitudes. Communities inside the Arctic Circle, located at 66.3°N, watch the Sun dip below the horizon in mid-November and don't see our star again for another two to three months. The long days and nights last even longer closer to the North Pole. The geographic pole itself experiences six months of daylight and six months of nighttime, with the Sun rising and setting on the vernal (spring) and autumnal equinoxes.

The long dark of winter isn't completely dark—at least not at first. Once the Sun dips below the horizon for the last time until the new year, the skies in the Arctic Circle seem to linger in a state of everlasting sunset. Blue skies take days to fade to orange, then a few more days to taper to red. Eventually, though, the color wears off, and a weeks-long twilight begins as the Sun falls deeper below the horizon and light slips ever farther out of reach. Complete darkness envelops regions only above 88°N. South of there, just enough light peeks through the upper atmosphere to cast a faint blue glow during the day.

## Dark Leads to Cold

Months without any meaningful sunshine take quite the toll on the local climate. Temperatures often drop quickly at night due to radiational cooling. Not long after sunset, the land begins losing the heat it accumulated during the day, radiating its warmth into the atmosphere. This process unfolds in abundance once the Sun dips below the horizon for months at a time.

It's tough for the Sun to heat up the polar regions even during times of daylight. Much of the land here is usually covered in snow and ice. These white surfaces have a high albedo, which means they reflect sunlight very efficiently. If you've ever found yourself squinting while shoveling snow on a sunny day, you've experienced snow's high albedo firsthand. This causes solar radiation to reflect back into the atmosphere, preventing the ground from absorbing much of its warmth.

Despite the fact that both the Arctic and Antarctic experience the same months-long night during the winter, the two polar climes have different types of weather. Antarctica gets much colder than the Arctic during the heart of winter because Antarctica is a continent. Land has lower heat capacity than water, so land warms up quickly and loses its heat just as fast at night. The Amundsen-Scott South Pole Station, a permanent settlement for scientists mere feet from the South Pole, routinely experiences temperatures below -60°F during the winter months.

Most of the Arctic Circle consists of the Arctic Ocean. Even though the water gets cold enough to support a thick crust of ice, water's ability to retain its heat doesn't allow air over the Arctic proper to get as cold as it does in the Antarctic.

While the North Pole itself doesn't get as cold as the South Pole, the far northern latitudes have a lot of land that gets plenty frigid during the winter months. Alaska, northern Canada, and the vast expanse of Siberia in central Russia can get awfully cold. An average day in the middle of January in Fairbanks, Alaska, features a high temperature of 0°F with a low temperature clocking in at a balmy -17°F. Temperatures frequently get even colder in these regions, where it's not uncommon for lows to drop below -40°F on a particularly chilly day.

All that polar air doesn't stay in the Arctic. Low-pressure systems that form in the polar regions can send cold fronts plunging into the lower latitudes, plummeting temperatures far below normal for thousands of miles away from the Arctic. Freezing air can even sink as far south as typically tropical Florida, damaging orange crops and causing cold-blooded lizards to fall out of trees as their bodies seize up to protect themselves from the chilly air.

# NOR'EASTERS, BLIZZARDS, AND LAKE EFFECT SNOW

**F**ew weather events command the full and undivided attention of millions of people quite as well as the threat of an impending snowstorm. Schools close, travelers fret, and shoppers hoard all the bread and milk they can find. Meteorologists field hundreds of panicked messages demanding to know exactly how much snow will fall. The storm itself garners wall-to-wall coverage on the news from before the first flake until the last pile is scooped up, making these rather routine winter events feel like the beginning of a war.

While so much of the reaction to wintry precipitation can feel like a heaping dose of melodrama, snowstorms are serious business no matter where they unfold. This holds especially true in the United States, where a huge number of people rely on miles-long commutes to get to work, school, and the grocery store. A big-time snowfall in a major metropolitan area can seriously disrupt the day-to-day lives of folks in the storm's path and cause wide-ranging effects that ripple across the rest of the country.

## Some Snow Stands Out

Snow and ice are diverse. There's a whole spectrum of storms that can produce snow, sleet, and freezing rain, ranging from small showers that put down a quick dusting to record-breakers that immobilize entire cities for a week or longer. Light snow showers might lead to traffic snarls on occasion, but a quick inch or two of snow in an area used to wintry weather isn't a big deal.

It's when a big storm comes through that the problems pile up. These noteworthy winter storms are usually heavy snow and ice falling on the cold side of a low-pressure system. Many dozens of winter storms trot across the United States every year. Swirling lows, which from above resemble a cinnamon bun, roll into the Pacific Northwest and bury the mountaintops in powder many feet deep. Small systems develop in southern Canada and race southeast across the United States. Intense lows explode over the Gulf of Mexico and race north to produce crushing snows over the Eastern Seaboard, coating thousands of square miles of densely populated land.

Most of these storms come and go without much fanfare, producing a blanket of snow that's tamed before the end of the day. But the most intense snowstorms can drop several inches of snow an hour, producing many feet of accumulation over the course of a day.

## Blizzards

The first word that comes to mind when we hear about big snowstorms is "blizzard." Blizzards have a strict definition, yet the term has taken on a life of its own over the years. It's similar to the way folks use "monsoon"—the rainy season in places like India and the American Southwest—to describe any torrential downpour of heavy rain.

A blizzard isn't just a big snowstorm or spate of heavy snow. The US National Weather Service considers blizzard conditions to occur when winds of 35 miles per hour cause blowing snow that drops visibility down to one-quarter of a mile or less for three consecutive hours. Those are true whiteout conditions, the kind during which the snow is so thick that you can barely see across the street in front of you. Someone caught outside during a blizzard can easily become disoriented, even in their own yard, which could lead to serious injury or worse if they can't feel their way to safety in a hurry.

Even though storms big enough to produce several feet of snow are usually powerful enough to generate the strong winds necessary for blizzard conditions, it doesn't always take a whole heap of snow to lead to a true blizzard. As little as an inch of dry, powdery snow blowing around in gusty winds is all it takes to lower visibility enough to pose a threat to motorists and folks trying to walk and work in the elements. Gusty winds blowing over loose snowpack can even lead to a ground blizzard, which can take place beneath otherwise clear skies.

# COLD AIR DAMMING

Winter storms aren't always a snowfall bonanza. The mountains play a key role in bringing plenty of freezing rain and sleet to lower elevations, through what is known as cold air damming, even as many surrounding communities enjoy a snowy day.

Some of the most dramatic cold air damming in the world takes place over the Mid-Atlantic in the eastern United States. Cold northeasterly winds blowing over Virginia, North Carolina, South Carolina, and Georgia eventually run up against the Appalachian Mountains. Since the cold air can't blow around the mountains and the air is too heavy to simply flow over the peaks, the Appalachians act like a dam that forces all the cold air to pool up in a region called the Piedmont, or the foothills.

When a low-pressure system approaches the Mid-Atlantic and its circulation leads to cold air damming, warm southerly winds a few thousand feet above the surface can ride over the bubble of cold air at the surface. This setup causes snowflakes to melt as they fall into the warm air and refreeze as they enter subfreezing temperatures wedged down at the surface. As a result, cold air damming during a winter storm can lead to destructive ice storms in the Piedmont region.

# MOUNTAIN SNOWS

Nor'easters and lake effect snow are both impressive, but nowhere else in the world can do snow like the mountains. The peaks of tall mountains bear witness to dozens of feet of snow every year. Frigid temperatures at high altitude allow snow to fall on mountaintops year-round in some areas. Parts of the northern Rockies are no stranger to midsummer snow showers when a potent system moves overhead.

Mountains are especially susceptible to heavy snows as a result of upsloping, the same process that leads to the rain shadow effect in central Washington. Winds riding up the slopes of tall mountains can condense almost all of the moisture in the air, leading to persistent heavy snows that drop foot after foot of powder.

## Nor'easters

Not all big snowstorms are blizzards, but many blizzards take place during particularly large snowstorms. There are fewer snowstorms bigger or more impactful than nor'easters. These coastal storms plague the East Coast of the United States at least a handful of times each winter, coming around more often during active years and hardly at all during a warm winter.

Nor'easters are large low-pressure systems that form just off the coast of the eastern United States. A potent nor'easter with plenty of cold air to work with can pelt the region with heavy snow, damaging winds, and significant coastal flooding as it strengthens and moves north through the northwestern Atlantic Ocean.

The term "nor'easter" refers to the northeasterly winds that howl across coastal communities as these storms crank up. It's not uncommon for folks on Massachusetts's Cape Cod, for instance, to experience wind gusts well above hurricane force as the heart of a nor'easter roars through. These winds are the embodiment of the intense dynamics necessary to spawn and maintain these systems.

There's something truly special about nor'easters that sets them apart from other storms. Nor'easters have a touch of the ocean in them. The Gulf Stream, a warm current of water that flows northward from the Gulf of Mexico parallel to North America's Eastern Seaboard, can provide a boost to nor'easters as they organize over the western Atlantic Ocean. This warmth can help one of these coastal systems strengthen faster than it would have alone through the strong jet stream above.

A low-pressure system that strengthens at a tremendous pace goes through a process called bombogenesis. A storm that bombs out, so to speak, experiences a drop in its central minimum pressure of at least 24 millibars in 24 hours. Nor'easters that undergo bombogenesis are likelier to produce the exceptional snowfall rates and driving winds that make these storms so memorable.

The ocean-dwelling nature of nor'easters also opens the door for deep tropical moisture to seep into the storm from the south. This moisture provides a boost to precipitation in the storm, encouraging snow totals that are measured in feet instead of inches.

One of the most striking features of a classic nor'easter is the beautiful comma-head shape that gives these storms such a gorgeous appearance on satellite imagery. The heaviest snowfall within a nor'easter typically occurs within the curly appendage on the northwestern side of the storm. The intense winds that spiral around a nor'easter in the middle and upper levels of the atmosphere eventually run up against the environment around them.

These fast-moving winds press against slower air outside of the storm, creating a horizontal stretching motion throughout the atmosphere. This area is called a deformation zone. This stretching motion causes air to rapidly rise from the surface, fueling thick bands that can dump several inches of snow in an hour. Historic nor'easters that swept over the megalopolis—the stretch of densely populated cities along Interstate 95 between Washington, DC, and Boston, Massachusetts—saw this deformation zone form and move right over those cities.

An infamous storm nicknamed the "Superstorm of 1993" formed in a near-ideal environment that brought accumulating snow as far south as New Orleans, Louisiana. The system's intense deformation zone produced double-digit snowfall totals from central Alabama through southern Canada, leaving behind an all-time record measurement of 13 inches of snow in Birmingham, Alabama.

## Lake Effect Snow

Cities along the Atlantic Ocean don't hold a monopoly on jaw-dropping snowstorms. The Great Lakes of North America, including Lakes Erie, Huron, Michigan, Ontario, and Superior, are famous for the bands of snow that can let loose and rip across communities unlucky enough to find themselves downwind of these relatively warm bodies of water.

Lake effect snow is one of the world's most impressive weather phenomena. Bands of lake effect snow build through convection, the same process that allows thunderstorms to blossom on a humid day. The Great Lakes are big, deep, and they're pretty far north. This puts them in the perfect position to experience intense cold snaps long before the surface of the lakes freezes over. Residents of waterfront cities like Chicago and Cleveland are familiar with the lakes' resistance to fast temperature changes during summer heat waves, when neighborhoods near the lake stay cool due to the water's influence on the air above it and during the winter, when those same communities stay slightly warmer than their landlocked neighbors.

If cold air combines with sufficient moisture and blows over the unfrozen lakes, the warm water can heat up the air directly above the surface. This warmed air quickly rises through the cold air above, causing bands of snow to develop. Winds organize

The temperature contrast between cold winds blowing over a lake's warm waters can spark small bands of snow showers that often drop heavy accumulations on the lakeshore.

RISING AIR

COLD WIND

SNOW↓

WARMER WATER

these bursts of heavy snow into long bands and blow them ashore. The process can continue unabated for days until winds shift, temperatures change, or the air dries out.

It's astounding how much snow can fall during a substantial lake effect snowstorm. Watertown, New York, which lies on the eastern shore of Lake Ontario, averages about 100 inches of snow every year. Buffalo, New York, once received 82 inches of snow (that's nearly seven feet!) over the course of five days during a persistent lake effect snow event in late December 2001.

The Great Lakes aren't the only place that can experience lake effect snow. Any sizable lake or bay can generate bands of snow if cold, moist air flows over water that's relatively warm in comparison. The same process is possible over parts of the ocean, as well. Some of the world's heaviest snow falls on Japan's northern island of Honshu. Cold air sweeping across China and Russia flows southeast over the warm Sea of Japan and generates convection that pours feet of snow on the island. It's this oceanic effect snow, as they call it, that makes Sapporo, home to more than two million people, one of the snowiest major cities in the world.

# POLAR VORTEX

Polar night in the Arctic Circle creates some of the most unforgiving conditions on Earth.

Once the winter sun dips below the horizon for the last time, a frigid chill settles over the tundra for months on end. Temperatures can drop below -60°F during the bitterest cold snaps in Siberia and interior Alaska. It takes a strong belt of winds, known as the polar vortex, to keep the season's coldest air confined to the top of the world. However, turbulence in this ever-present feature can lead to some of the nastiest spells of winter in the lower latitudes.

The polar vortex is a jet stream that encircles the Arctic and Antarctica during the cold seasons. Meteorologists often compare it to a moat or a fence that confines bitterly cold air to the poles during the dead of winter. The one over Antarctica is of less concern to humans because any shifts happen over the ocean. However, shifts in the polar vortex in the Arctic Circle can mean huge shifts in temperature in the Northern Hemisphere.

# POSITIVE
## ARCTIC OSCILLATION

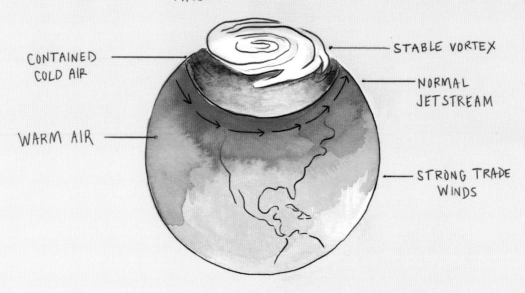

CONTAINED
COLD AIR

STABLE VORTEX

NORMAL
JETSTREAM

WARM AIR

STRONG TRADE
WINDS

# NEGATIVE
## ARCTIC OSCILLATION

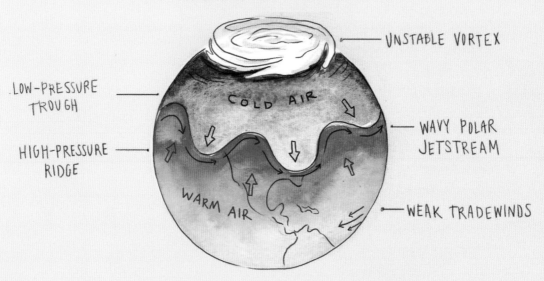

UNSTABLE VORTEX

LOW-PRESSURE
TROUGH

COLD AIR

HIGH-PRESSURE
RIDGE

WAVY POLAR
JETSTREAM

WARM AIR

WEAK TRADEWINDS

Scientists can measure the stability of the polar vortex using an index called the Arctic Oscillation (AO). A normal winter day features a positive AO, which indicates a strong, smooth, and stable polar vortex. That means that the belt of winds is tightly wrapped around the Arctic Circle, ensuring the coldest cold stays north and the weather in the rest of the Northern Hemisphere remains relatively benign. Things start to get dicey when a wavy jet stream disrupts the polar vortex. A powerful storm system or a steep ridge in the jet stream can set the polar vortex's circulation off-balance, much like a spinning top wobbling out of control when it hits a bump.

The ensuing waviness in the polar vortex circulation—or a negative AO—often sends big ridges of higher pressure into the Arctic and steep troughs of lower pressure toward the lower latitudes, forcing warmer air to jut north and frigid air to dive south. This is often the driving force behind the weather patterns that allow subzero temperatures to bathe the Great Plains as temperatures climb to near freezing at the North Pole in the dead of winter. These cold troughs can also help create strong low-pressure systems at the surface, allowing major snowstorms to track across heavily populated parts of the United States and Canada.

Sometimes, though, the circulation grows so unstable that an entire piece of the polar vortex can break off in what is known as a cutoff low. A cutoff low is a center of low pressure in the upper levels of the atmosphere that has completely broken free from

# SNOW IN MIAMI

Southern Florida is famously immune to the throes of winter that often envelop the rest of the United States. The state is far enough south—and surrounded by enough water—that even the snappiest cold is relatively toothless by the time it reaches this swampy terrain. A lobe of the polar vortex dipped over Atlantic Canada on January 19, 1977, plunging cold air all the way to Cuba. Temperatures dipped below freezing in Miami just long enough for rain to briefly change over to snow. It was Miami's first and only instance of snow on record, an event so remarkable that local newspapers headlined the unprecedented wintry precipitation in a font size that rivaled the announcement of the Moon landing.

# SUNDOGS

An extended cold snap provides a wonderful opportunity to see dazzling lights appear around the Sun. Most high-level clouds on a frosty day are made up of tiny ice crystals. These ice crystals can refract light at a certain angle, allowing halos and rainbows to appear around the Sun if you're standing in the right spot at the right time.

The most common type of halo to see on a cold day is a sundog, or a bright and colorful patch of light that appears at about 22° on either side of the Sun. Sundogs are visible anywhere on the Earth when sunlight shines through ice crystals in cirrus or cirrostratus clouds, but you're more likely to see them at higher, colder latitudes and when the Sun is low on the horizon. Sundogs take their name from the colorful phenomenon's resemblance to loyal dogs sticking by the Sun's side on a chilly afternoon.

the jet stream. There's no flow to steer it along, so the cutoff low just meanders until another trough dips south and scoops it back to higher latitudes. These features have a nasty habit of lingering for many days at a time, bringing with them some of the coldest temperatures possible outside of the Arctic. Historic cold snaps in 1985, 1996, and 2014 were all the result of cutoff lows that originated in the polar vortex.

The polar vortex turned into a household term in 2014 when a tremendous chill descended over Canada and the United States, shattering daily record low temperatures as far south as the Gulf of Mexico. The intensity of the cold wasn't as noteworthy as its duration. This event wasn't just one or two cold mornings—it lasted an entire week. Chicago, Illinois, saw subzero low temperatures on seven out of eight mornings between January 2 and January 9, falling as low as -16°F on January 6.

The bitter temperatures of early January 2014 were far from the lowest ever recorded, but it was the first intense cold snap to strike since social media came into widespread use. Meteorologists discussed the polar vortex in their forecasts, and the delightfully frightful term took the internet by storm. While it sounded new and terrifying to folks who aren't avid weather followers, the first use of polar vortex appeared in an 1853 edition of *Littell's Living Age,* which served as a sort of *Reader's Digest* of its time. In an essay steeped in the era's grandiose language, the author described how warm winds flowing from the equator ultimately met their destiny against "the great polar vortex," and so became the term we use today.

# AURORAS and SOLAR STORMS

**W**hile light may fade after sunset, the Sun's influence lasts well after darkness settles over a frigid wintry landscape. The northern lights are one of the most impressive sights on Earth. Witnessing a theatrical display of colors dancing through the night sky usually tops the must-do list for any winter-hardy nature lover. These natural lights high in Earth's atmosphere are as much a product of space as they are a product of Earth itself.

# Earth Is Magnetic

The beautiful lights that wave and wiggle above begin with the colossal ocean of molten metal that churns thousands of miles beneath our feet. Earth's solid inner core is surrounded by a hot liquid outer core that consists of iron and nickel. The temperature of Earth's core approaches 10,000°F. The blazing heat from the inner core heats up the outer core, constantly churning about the liquid metal as a result of convection.

That deep reserve of liquid metals sloshing around in the center of Earth induces an electric current that generates a powerful magnetic field around the planet. Earth's magnetic field envelops the planet like a large, invisible bubble, protecting the planet and its atmosphere from the harsh conditions of outer space.

If you could see Earth's magnetic field from space, its outline would look something like a cross section of an onion. The field would bulge out and peel away at the sides, arching down toward Earth's surface at the magnetic north pole and magnetic south pole. The layered shape of this magnetosphere deflects harmful solar radiation and causes some fascinating interactions with our atmosphere.

The magnetic poles aren't the same as what we picture when we talk about the North and South Poles. Earth has true poles and magnetic poles. The true North and South Poles are fixed geographic locations, where Santa keeps his workshop in the Arctic and explorers pose for photographs on Antarctica. True north and true south line up with the planet's axis, serving as the top and bottom of the world when you look at Earth relative to its rotation.

There's also magnetic north and magnetic south, which are several hundred miles displaced from the true poles. The distance between true north and magnetic north is the magnetic declination, and calculating the angle between your position relative to these two points on Earth is a crucial piece of information for anyone who relies on a magnetic compass for direction.

Earth's magnetic field isn't static. The orientation of our planet's magnetism is always slowly on the move. The magnetic north pole has moved hundreds of miles over the last couple of centuries, shifting from northern Canada toward northern Russia since the 1990s.

## Solar Wind

We're about 93 million miles from the Sun, and we don't need to get any closer to know that it's hot. Really hot. The star of our solar system burns at an average temperature of more than 1,000,000°F just above its surface, the effects of which we can readily experience when we step outside on a blazing summer afternoon. The Sun is powered by intense nuclear fusion at its core. The immense heat generated by this chain reaction radiates toward the outer layer of the Sun, known as the photosphere, where gases like hydrogen and helium burn bright as a hot plasma.

In addition to the photosphere that emits visible and infrared radiation, the Sun also has an atmosphere that plays a significant role in the star's interaction with the Earth. The most prominent part of this solar atmosphere is the corona. We can actually see the corona during a total solar eclipse. When the Moon completely covers the Sun and the eclipse reaches totality, observers can see a faint, hazy ring emanating from the Sun. This fuzziness is plasma within the Sun's corona.

The immense heat within the corona causes the Sun's atmosphere to emit a steady stream of charged particles and electromagnetic radiation in all directions. This constant flow of charged plasma away from the Sun is the solar wind. This solar wind constantly washes over Earth's magnetosphere, causing the shape of the magnetic field around our planet to deform a bit, swooping back downstream from the solar wind.

## Auroras

Plasma and radiation within the solar wind funnel down toward the top of the atmosphere near the magnetic north and south poles. This flood of charged particles ionizes gases in the upper atmosphere, generating bright lights that sparkle high in the night sky over polar regions of the world. These lights are auroras.

Auroras have different names in each hemisphere. The aurora borealis occurs in the Northern Hemisphere over the Arctic regions, while the aurora australis paints skies in the Southern Hemisphere over the Antarctic. Given the distribution of land and population around the world, lots of people live around the Arctic Circle, and hardly anyone lives near the Antarctic Circle. As such, the northern term "aurora borealis" has become synonymous with the aurora itself. No matter whether you call them by their formal names, the northern lights, or just an aurora, it's all the same beautiful phenomenon.

You're more likely to see an aurora if you're a fan of the cold weather and wish to visit polar regions during the winter months. Auroras are almost always present day and night regardless of the season, but the long nights of winter make these formations readily visible to the naked eye without getting obscured by pesky sunlight.

An aurora can appear different colors depending on which gases the charged particles are ionizing in the upper atmosphere. Ionized oxygen and nitrogen particles generate vivid reds, blues, and yellows. Combinations of these primary colors result in the oranges, greens, and purples that so often tint these fabulous displays.

The Sun has a magnetic field as well. Imbalances within the magnetic field can allow the surface of the Sun to eject huge amounts of plasma and electromagnetic radiation into space. These emissions are solar flares. If a solar flare is directed toward Earth, the sharp increase in the density and speed of the solar wind can greatly increase the intensity and brightness of auroras. Particularly intense solar flares can bombard Earth with so much energy that auroras reach down to the middle latitudes, sometimes tinting the sky all the way down to the southern United States.

## Solar Storms

The effects of intense solar winds don't end with auroras. A coronal mass ejection (CME) is an intense emission of plasma and electromagnetic radiation from the Sun's corona. Closely associated with solar flares, CMEs directed toward Earth can wreak havoc on communications networks and electrical grids if the emission is strong enough.

A powerful solar storm in 1989 damaged an electrical grid that powered much of Quebec, causing an extended power outage in the Canadian province. One of the most intense solar storms on record occurred in the mid-1800s, generating brilliant auroras and even shorting out some telegraph equipment in the northeastern United States.

The potential risks associated with solar storms make it imperative for scientists to keep track of the Sun's every move. While the latest generation of GOES weather satellites is valued for the high-definition data the updated instruments provide of the Earth's atmosphere, the back of the satellite, which faces the Sun, contains instruments that monitor the solar surface for signs of activity. The Solar Ultraviolet Imager generates detailed views of the Sun that can help scientists look out for solar flares that could cause trouble here on the Earth.

# OPTICAL ODDITIES: HALOS, STARS, and PILLARS

Nature doesn't sit still. Rising air drags towering thunderstorms to the edge of the atmosphere. Raging downpours send torrents of water cascading through city streets. Spiraling lows blanket entire continents with clouds woven with artistic intricacy. But some of our world's most exciting sights occur when it seems like there's not much going on at all. From starlight that originated in faraway galaxies to delicate rays of sun shining through the clouds, the simple act of light passing through the air and moisture above provides scenes that are a joy to witness.

## Twinkling Stars

Singing your way through "Twinkle, Twinkle, Little Star" is an inescapable part of childhood. It's one of life's little pleasures to stand outside on a clear evening and take a few moments to stare up at the stars as they flicker in the dark sky. The sparkling effect is especially charming for folks who grew up in light-polluted cities and suburbs and venture into a rural area at night and take in thousands of stars for the first time.

Even though it looks as if the stars are pulsating like strobe lights, they're not really twinkling at all. Light from space has to filter through every layer of the atmosphere on its way down to our eyes. Some layers are colder than others, and there are alternating pockets of bone-dry air and high humidity that stack up from the surface all the way to the edge of space.

The density of air changes slightly depending on its temperature and humidity, and strong winds aloft force these different layers to constantly move around and mix with one another. As a result of all that churning air, light doesn't move in a straight line as it travels toward the surface. The slight turbulence in the path of a ray of light passing through the fluid air above leads to the twinkling effect we see when we gaze upon the stars.

## Halos

Light does more than just shimmer when it collides with the countless particles floating around above our heads. Ice crystals hanging high in the sky do a great job scattering sunlight in spectacular ways. The shape and thickness of the ice crystals that make up high-flying clouds determine how sunlight and moonlight bend through these frozen formations.

One of the most striking interactions between light and moisture in the air is a brilliant halo. A halo is a bright light that appears to encircle the Sun or the Moon when there's a thin layer of ice crystals or cirrus clouds hanging high in the sky. Sometimes, the layer of ice crystals is too thin to see them from the ground, but it's just thick enough to bend sunlight or moonlight into an unmistakable halo.

The most common halo is called a 22° halo, formed when sunlight or moonlight refracts, or changes direction, as it passes through a long, hexagonal ice crystal that's oriented parallel to the ground. Sunlight hits the top of the ice crystal and bends 22° as

## BIG MOON, GULLIBLE BRAIN

Have you ever happened upon a Moon that looks enormous on the horizon? Our orbital companion can look so large that it almost seems like it's falling out of the sky. However, this isn't an optical phenomenon—it's an optical illusion! We're used to seeing the Moon high in the sky, far away from objects like trees and buildings. When the Moon is low on the horizon, however, it looks enormous compared to our relatively tiny surroundings here on the Earth. The perspective of tiny objects against a huge Moon tricks our brain into making the Moon look much larger when it's low on the horizon.

it passes through the formation, focusing the light into a tight, bright halo that surrounds the Sun or the Moon. Depending on the size and the orientation of the ice crystals, a much wider 46° halo is also possible.

While vivid halos are relatively common around both the Sun and the Moon, the Sun's intensity leads to halos that appear much more brilliant and more diverse in the sky. A halo itself is only the beginning. Sunlight passing through flat ice crystals can lead to tangent arcs, or partial halos that appear above, below, and to the sides of the Sun.

A circumzenithal arc can appear as a bright upside-down arch above the Sun. A parhelic circle appears as a thin belt of light that emanates horizontally across the entire sky at the same altitude as the Sun. Circumhorizontal arcs, such as sundogs, show up on either side of the Sun. Some circumhorizontal arcs are so broad that they cast a vivid rainbow across the clouds in which they form. These fire rainbows, as they're often called, are elegant enough to make you stop and gawk at the emerging display of colors.

# SPOT THE SATELLITES AT NIGHT

Stars and planets aren't the only lights we see in space at night. It's possible to spot artificial satellites and even the International Space Station if you look in the right place at the right moment.

We can see human-launched objects in space because they orbit quite high above the surface. The International Space Station orbits at an altitude of nearly 250 miles, which means that the station is still in full view of the Sun even though it's long after sunset on the ground below. The solar arrays and shiny metallic exteriors catch the sunlight and reflect it back to the surface, allowing us to see these bright objects zip across the night sky.

Tangent arcs are prolific in northern latitudes during the colder months when layers upon layers of icy clouds fill the sky, but folks as far south as Florida can enjoy halos with tangent arcs if upper-level temperatures are cold enough to support icy clouds.

# CREPUSCULAR RAYS

Even a scant amount of light is all it takes to provide a memorable moment in a sky full of clouds. Sunlight poking through a break in thick cloud cover creates a crepuscular ray—a column of light reaching to the ground. Just as we can see sunshine beaming through a window into a dusty or smoky room, crepuscular rays are visible because they're shining through moisture or haze hanging around in the air below the clouds.

Anticrepuscular rays, the opposite phenomenon, are possible when clouds cast a long shadow across an otherwise clear sky. It's common to see anticrepuscular rays in flat areas like the Great Plains when there are thunderstorms on the western horizon around sunset. The tall tops of the storms can block sunlight for dozens of miles, casting a narrow shadow overhead.

## Coronas

When light shines through a thin layer of clouds made up of water droplets instead of ice crystals, a bright splotch or ring of light can appear on the clouds as if a spotlight were shone on the clouds from above. This effect, called a corona, forms when light diffracts around a water droplet rather than refracting through an ice crystal, like the process that creates a halo.

A corona is more apparent at night than during the day because the effect is muted amid the bright sunshine but prominent when it fills much of the night sky. If you ever look up at the Moon on a cloudy night and it looks like it's being seen through a bright, blurry film, that's very likely a corona.

## Cloud Iridescence

The edges of a corona around the Moon are sometimes tinted shades of red or brown. This is an example of cloud iridescence. The phenomenon is most prevalent during the day when sunlight shines through a thin layer of clouds made up of water droplets of different sizes, appearing as a faint palette of pastels within the clouds. Cloud iridescence is especially common near the tops of thunderstorms when the upper edge of the clouds just begins to obscure the Sun. Iridescence is the same effect we see when we look at an oil sheen in a dirty parking lot or the surface of a soap bubble when it catches the light just right.

## Sun/Light Pillars

Not all light is spread out and twisted apart as it interacts with water and ice drifting through the sky. The low angle of the Sun in the sky around sunrise and sunset can cast a pillar of light on the clouds that appear above our bright star. These sun pillars are created by sunlight reflecting off of flat ice crystals, leaving a long streak of light in the sky. Sun pillars are reminiscent of city lights shining off the surface of a calm bay.

Exceptionally cold climates that support the formation of ice crystals near the ground can cause light pillars to appear above artificial lights as well, lending a strange aura to Arctic towns as thin columns of light shine above streetlamps, traffic lights, and homes.

# SPRING

Centuries ago, folks defined the transitions between summer and winter by the plants. The end of summer saw the "fall of the leaf," while the end of winter saw plants and leaves spring forth into existence, and so the names for the two transition seasons were born. Spring is the slow transition from the chill of winter to the steam of summer. Building heat allows intense thunderstorms to crescendo through the spring months, providing breathtaking scenery when they're harmless but sometimes leaving widespread damage and tragedy in their wake.

# VERNAL EQUINOX

**S**pring is a time for growth around us and above us. Days grow longer, temperatures grow warmer, and thunderstorms grow taller in the sky. It's a fascinating time of the year for folks who get a thrill from extreme weather. Strong storms produce lightning hotter than the surface of the Sun, and mile-wide tornadoes can emerge from a gentle breeze. It's almost hard to imagine the same environment that sparks such powerful storms can sustain delicate fields of flowers and the crops that feed us all, but that's what makes spring such an alluring time of year for those of us who treasure everything nature can throw our way.

## The Equinox

The vernal equinox is one of the two days each year that the Sun's radiation shines directly on the equator. It's a common belief that on the two equinoxes, both vernal and autumnal, we experience equal amounts of daylight and nighttime around the world. It doesn't quite work out that way. The date of the equilux, as it's called when day and night are of equal length, varies because of the properties of the Sun and our atmosphere.

Sunrise and sunset are when the geographical center of the Sun touches the horizon. Since the Sun looms large in the sky, we begin receiving light from the top of the Sun before sunrise and we experience daylight until the top of the Sun slides behind the horizon at the end of the day. This phenomenon, combined with the fact that the Sun's direct rays cross the equator only at a precise moment, means that the day is always a little bit longer than precisely 12 hours near the equinox.

For example, the vernal equinox in 2021 fell on March 20. Los Angeles, California, witnessed its equilux on March 16 that year, basking in sunshine for 12 hours and 15 seconds that day. A few thousand miles to the northeast, Toronto, Canada, saw only 11 hours and 57 minutes of daylight on the same day. Canada's largest city didn't meet its equilux until March 17, when Torontonians saw 12 hours and 58 seconds of daylight between sunrise and sunset.

## Climate and the Calendar

The length of the day isn't the only thing that is a little amorphous. So is the very concept of spring itself. The vernal equinox is the first day of astronomical spring, the point when the Sun's direct rays slowly start to heat the hemisphere on the long march to summer. While we consider the equinox the first day of spring, meteorologists and climatologists adhere to the calendar to define the seasons.

> In the spring, I have counted 136 different kinds of weather inside of 24 hours.
>
> —MARK TWAIN, speech, 1876

Climatological seasons—sometimes called meteorological seasons as well—break down the year into four seasons based on the Julian calendar. Climatological spring begins on March 1 and lasts through May 31. Not only are these seasons easier to keep track of than the precise date of the equinoxes and solstices, but it makes it much easier for scientists to keep track of seasonal weather data. Often, in regions in the middle latitudes, you find that weather changes line up better with climatological seasons than with astronomical seasons.

Those changing weather patterns are no joke. After the vernal equinox, increasing solar radiation bathes more of the Northern Hemisphere with each passing day. This barrage of warmth slowly heats the environment from south to north, forcing abrupt changes in weather patterns.

The sharp temperature gradient from the warming southern latitudes and the still-frosty northern latitudes enhances the speed of the jet stream, allowing this upper-level band of winds to move farther to the south. A stronger and wavier jet stream allows potent low-pressure systems to develop and let loose a whole barrage of hazardous weather conditions, ranging from days-long windstorms to tornadoes that can cut a path through the heart of a city in minutes.

While these shifting weather patterns are fodder for storm chasers who stalk the intersection of winter and spring in search of hulking thunderstorms, the first tangible change on the journey into the warmer months is the rebirth of the plants that cover and color our world.

## Frosts and Freezes

While climatological spring begins on March 1 and astronomical spring begins three weeks later, the date on the calendar or the position of the Sun in the sky doesn't mean much if it's still too cold to plant your garden. Tracking the last frost and freeze of the season is an objective measurement of when winter finally gives way to spring.

The frost and freeze are important factors for both commuters who are late for work (scraping windows takes time!) and for folks who maintain gardens and tend crops. Frost is a thin layer of ice that forms on exposed surfaces on a cold night. The actual air temperature doesn't have to drop below freezing for frost to develop. A temperature below 36°F is sufficient for the air directly above the surface to drop to or below freezing, allowing that thin layer of frost to form. Some plants, like succulents and houseplants left on patios and balconies, are susceptible to frost damage, but this is preventable by gently covering plants with tarps or blankets to fend off the morning's cold.

A freeze, aptly enough, occurs when the air temperature reaches 32°F or lower. Many plants will bounce back from a light freeze, which sees temperatures as low as

# POLLEN STORMS

Beehives come to life during the spring months as bees collect pollen for themselves and the plants they visit. But bees aren't the season's only pollinators. The strong winds of spring transport massive amounts of pollen long distances, much to the chagrin of those of us who suffer from allergies. The first windy day after trees sprout their leaves is a brutal experience for folks whose immune systems consider pollen spores to be belligerent intruders. Blustery conditions ahead of a thunderstorm are particularly unpleasant when the gusts buffet freshly grown leaves ready to spew their pollen into the air, filling the sky with a yellowish-green haze that lowers visibility and leaves your eyes scratchy as you run for cover.

28°F, but any lower than that is a hard freeze, which can cause serious injury or even death in plants that can't handle the shock of low temperatures.

Given the tender nature of so many plants when it comes to cold temperatures, keeping track of the last freeze is serious business during the spring planting season. The average date of the last freeze varies widely across such a large and diverse country as the United States. The northern Gulf Coast typically feels its last freeze in early February, while communities near the border between the US and Canada can still dip below freezing into the middle of June or beyond. Areas in warmer states like California and Florida can go many years without ever experiencing subfreezing temperatures.

Average dates, of course, aren't a hard-and-fast rule. If the last freeze in the beautiful town of Podunk occurs on March 1 for five years and on March 31 for the next five years, the average date of the last freeze in Podunk would be March 16. Warmer years will see the last freeze occur much earlier in the spring than in a year when brutal cold snaps are the norm. However, the long-term data reveals an unsettling trend: cold temperatures aren't lasting as long into the spring as they used to. The retreating date of the last freeze is one of the most noticeable and tangible effects of climate change.

Climate Central, a nonprofit organization run by scientists and journalists who focus on climate change, conducted a study of weather data between 1970 and 2019 to determine the amount of warming across the United States during the spring months. Almost all of the United States, save for the northern plains around North Dakota and South Dakota, experienced warming of at least 1°F during the spring months, with the most dramatic warming occurring in the Southwest, Ohio Valley, and Mid-Atlantic states. As a result, the average date of the last freeze shifted earlier by as much as a week or longer in cities like Topeka, Kansas; Dayton, Ohio; and State College, Pennsylvania.

## Permafrost

Some areas never thaw out at all. Vast stretches of land in the northern latitudes don't receive enough solar radiation or warmth for the land to thaw out after a long and brutal winter. This land, called permafrost, stays frozen all year round. A gaze upon permafrost reveals a barren landscape devoid of the robust vegetation that covers so much of the rest of the world.

Areas near permafrost—land that freezes hard but thaws out during the warmer months—experience a fascinating interaction between the skies above and the ground below. Water expands and contracts as it freezes and thaws out. The pressure from freezing and thawing can warp the ground in dramatic fashion. We can see this in action every winter when we drive over gaping potholes and leap over giant cracks cleaving through sidewalks.

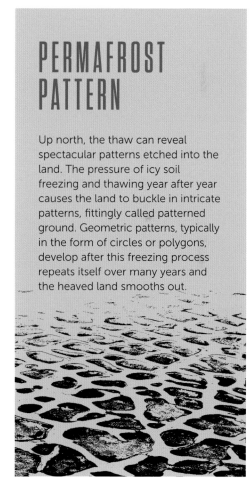

# PERMAFROST PATTERN

Up north, the thaw can reveal spectacular patterns etched into the land. The pressure of icy soil freezing and thawing year after year causes the land to buckle in intricate patterns, fittingly called patterned ground. Geometric patterns, typically in the form of circles or polygons, develop after this freezing process repeats itself over many years and the heaved land smooths out.

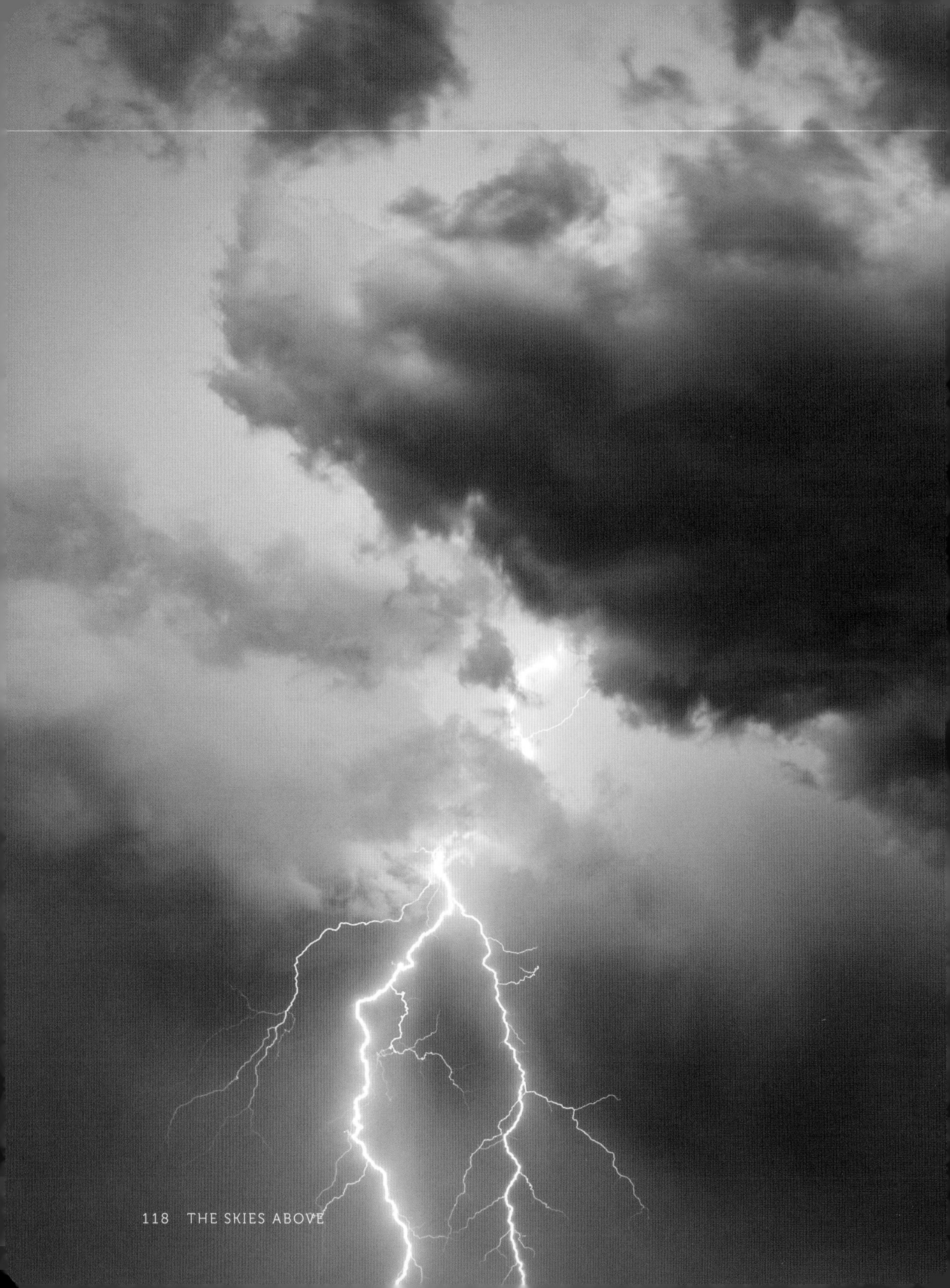

# THUNDERSTORMS
## AND LIGHTNING

**A** strong thunderstorm on a hot afternoon is an adventure for the senses. It's invigorating to breathe in the first crisp gust of wind scented by petrichor, an earthy smell, followed by a crescendo of thunder that precedes sheets of driving rain that wash away the grip of a steamy day. Thick, colorful bolts of lightning zigzag through the sky as they relieve the static field generated by the boisterous storm. Watching a thunderstorm run through its life cycle is even more awe-inspiring when you consider that each one, from a tiny shower to the fiercest torrent, begins with a single column of rising air.

# PETRICHOR

You really can smell the rain before it starts raining. Human noses are extremely sensitive to petrichor, the strong, earthy smell we detect during a spring or summer rain shower. The scent doesn't come from the rain itself—it's actually a chemical compound produced by bacteria in the soil called geosmin. Raindrops splashing on the soil aerosolize the geosmin and release it into the wind that blows ahead of the storm, signaling its arrival before the first drops fall.

## Convection

Imagine a big bubble of air just above Earth's surface. This bubble, or parcel, gets heated up during the day by the Sun, the ground, and thousands of buildings and vehicles that radiate the warmth they've absorbed on a sultry afternoon. Different types of ground cover lead to uneven heating of the air above the surface. After all, anyone who's ever burned their feet on a hot sidewalk knows the soothing relief of jumping onto the grass to cool their soles.

A parcel of air over an airport's mammoth parking lot gets far hotter under bright sunshine than a similar parcel of air over a patch of woods nearby. The uneven heating between the parking lot and the woods can allow the parcel of air over the pavement to begin rising, as its warmth makes the air less dense than all the air around it.

This rising parcel of air encounters a much cooler environment as it climbs hundreds and thousands of feet into the sky. The growing temperature difference between the warm, rising air and the cool air around it causes the warm air to ascend even faster. Eventually, the rising air in this newly formed updraft cools enough that its relative humidity reaches 100 percent and the water vapor inside condenses into a cloud. Billowing clouds follow the updraft through the atmosphere until the rising air finally reaches the same temperature as the environment around it, at which point the parcel reaches equilibrium and begins to sink back toward the ground.

This is convection, the process through which thunderstorms develop. Thunderstorms can range from a brief downpour over a single neighborhood to a squall line that snakes across entire states. Meteorologists used to refer to individual storms as "cells" because of the small-scale, closed nature of the convective loop that feeds a thunderstorm.

While that phrase has fallen out of use over the years, it lives on in the names of the three different types of thunderstorms: single-cells, multicells, and supercells.

## Single-Cell Thunderstorms

Single-cell thunderstorms have many different nicknames—pop-up storms, popcorn storms, pulse storms—but they all refer to a small thunderstorm that appears quickly and rains itself out. Single-cells are the most common type of thunderstorm in the world, with tens of thousands of little convective bursts forming every year from the equatorial tropics to the edges of the Arctic Circle.

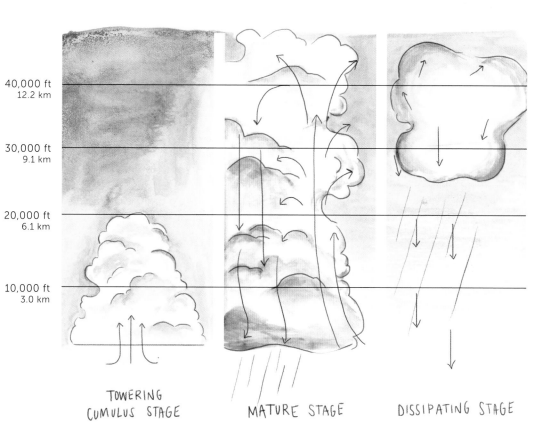

40,000 ft
12.2 km

30,000 ft
9.1 km

20,000 ft
6.1 km

10,000 ft
3.0 km

TOWERING CUMULUS STAGE    MATURE STAGE    DISSIPATING STAGE

The formation of a single-cell thunderstorm is simple and straightforward, which is why these quick spurts of rain are possible just about anywhere that gets warm enough to support convection. An updraft feeds unstable air into the thunderstorm until the weight of the raindrops suspended in the cloud is too much for the updraft to sustain. When the updraft can't support the weight of the water anymore, rain begins to fall out of the storm, and that precipitation drags cooler, more stable air down to the ground.

The flow of cool air out of a thunderstorm toward the ground is a downdraft. The downdraft hits the ground and creates an outflow, which is the refreshing wind you feel near a storm. That rush of stable air cuts off the inflow of unstable air into the thunderstorm. Surrounded by what's known as a cold pool, the thunderstorm runs out of unstable air and begins to wind down and dissipate.

The cold pool from one thunderstorm rolls along the surface like a miniature cold front, scooping up unstable air and forcing it to rise. This outflow can trigger the development of new thunderstorms nearby, which is why tropical climates are often drenched on a daily basis during the rainy season.

## Multicell Thunderstorms

If upper-level winds are blowing strongly enough, the thunderstorms that develop along that outflow boundary can merge with one another into a group or complex of storms. These complexes often move and evolve as if they were one large thunderstorm because they share a common root in that cold pool at the surface.

The most familiar multicell thunderstorms are squall lines. Sometimes called bow echoes when they form an arching shape on radar imagery, a squall line comes on suddenly and slowly tapers off as it passes overhead.

Strong, well-organized squall lines are able to survive for many hours and travel hundreds of miles before they dissipate. They accomplish this admirable feat thanks to that buildup of cold air under the storms. The outflow boundary beneath the line of thunderstorms doesn't force air to rise straight up into the sky. The updraft glides up and over the cold pool, feeding into the squall line at an angle. This allows the squall line to ingest warm air ahead of it and vent the cold air back and away from the heart of the storms. Squall lines can keep chugging along until they run out of unstable air or until their own outflow boundary runs so far ahead of the thunderstorms that it can't scoop air into the updraft anymore.

# SOME THUNDER IS LOUDER

The sound of thunder doesn't always depend on how close the storm is to your house. Air temperatures can also affect how loud a thunderstorm sounds. A temperature inversion, or a layer of warmer air sitting atop a layer of colder air near the surface, can act like a ceiling and deflect the sound of thunder back toward the ground. Loud claps of thunder from storms located near a temperature inversion can reverberate for many miles, making a thunderstorm's roar audible far away from where the lightning struck.

A squall line will make its presence known long before the wind hits. Shelf clouds are prominent, low-hanging clouds that can form along the leading edge of a thunderstorm as it races along its path. Air rising along a thunderstorm's outflow boundary condenses as it rounds the top of the pool of cold air, forming this pronounced shelf-like cloud that looms beneath the edge of an approaching storm. They aren't always a sign that severe winds are on the way, but the strongest squall lines often produce photogenic shelf clouds.

The most dangerous type of squall line is known as a derecho, which comes from the Spanish word for "straight." Derechos are relatively common in the central and southern United States during the warm season, as intense squall lines encounter near-perfect environments to strengthen and survive for a long time. The strongest derechos can start over the northern plains and race east until they emerge over the Atlantic Ocean. A derecho can produce devastating straight-line wind damage, blowing down trees, power lines, and even buildings. Wind gusts in the strongest derechos can exceed 100 miles per hour, leaving behind tornado-like damage across hundreds of communities.

## Supercell Thunderstorms

Winds don't always blow in the same direction throughout the atmosphere. Strong winds, for example, might blow from the south at 5,000 feet above the ground, from the southwest at 15,000 feet, and from the west at 28,000 feet above the ground. The change in wind speed and wind direction with height, known as wind shear, is responsible for the creation of a towering type of thunderstorm known as a supercell.

Powerful wind shear can lead to a horizontal rolling motion throughout the atmosphere. If a strong thunderstorm updraft rises through this area of wind shear, the updraft can bend that horizontal rotation vertically, forcing the entire updraft itself to begin spinning.

A rotating updraft makes a supercell thunderstorm exceptionally strong and resilient. The spiraling motion makes it difficult for upper-level wind shear to shred the thunderstorm apart. Instead, the shear tilts the updraft diagonally, causing the downdraft winds to blow away from the heart of the storm. This keeps the updraft free to ingest as much unstable air as it can consume.

A supercell thunderstorm almost resembles a miniature low-pressure system. The updraft itself acts like the center of the low, and the powerful downdraft winds blow around the thunderstorm like cold and warm fronts. The broad rotation around a supercell lends the storm a fish hook appearance on radar imagery, leading meteorologists to describe a classic supercell as a "hook echo."

When you see breathtaking photographs of a dark, chiseled thunderstorm over flat land, you're very likely looking at a supercell. These storms are unmistakable in their classic form. Supercells feature a vast anvil cloud that forms a wide, thin canopy when the updraft rises as far as it can and begins to spread out. A prominent wall cloud can hang low beneath the base of the storm as air condenses beneath the rotating updraft.

Sometimes, you can even spot the rotation of the updraft in the clouds, with clouds adorning the updraft as a spiraling column that stretches high into the sky. Tornadoes are picturesque when they jut from the clouds to the ground beneath a classic supercell, a striking contrast to the clear sky that often appears behind them on the other side of the storm.

## COUNT THE SECONDS BETWEEN LIGHTNING AND THUNDER

You don't always need a weather radar to tell you how far away lightning has struck. It takes about five seconds for the sound of thunder to travel one mile. When you see a flash of lightning, just start counting the seconds until you hear the thunder. If it takes 10 seconds before the rumble reaches you, the lightning struck about two miles away. No matter how far away the lightning is, it's important to head indoors the minute you hear thunder. If you're close enough to hear thunder, you're close enough to get struck by the lightning.

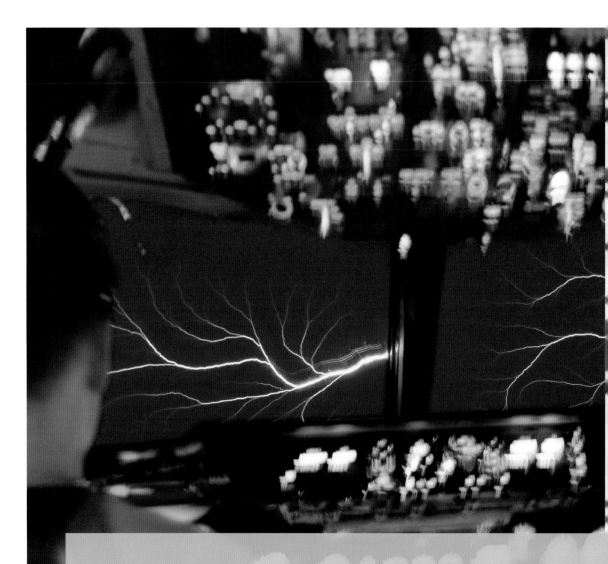

## ST. ELMO'S FIRE

A strong static buildup in the air can lead to more than lightning. The static can discharge as St. Elmo's fire, a faint ball of plasma around objects such as ship masts, telephone poles, and airplanes as they fly through building thunderstorms. The plasma appears as an eerie blue or violet glow around objects, sticking out into the electrical field built up in the air. St. Elmo's fire is largely harmless, but the static buildup required for its formation could mean lightning is imminent.

## Severe Thunderstorms

The vast majority of thunderstorms around the world are rather docile. They pop up, they rain for a little while, and they leave, soon to be forgotten. A handful of storms, though, pack such a punch that they can change the landscape itself. While all types of thunderstorms can be dangerous, supercells often produce the worst severe weather. The strongest supercells can create hailstones larger than baseballs and violent tornadoes that can grow more than one mile wide.

The term *severe thunderstorm* in the United States describes a storm that produces damaging wind gusts of 58 miles per hour or stronger, hail the size of quarters (one inch in diameter) or larger, or a tornado. During the 2010s, the United States averaged around 14,100 instances of damaging winds, 6,000 reports of large hail, and 1,300 tornadoes each year.

Damaging wind gusts are by far the biggest severe weather threat you're likely to encounter during a thunderstorm. Straight-line winds, which blow in one direction, as opposed to swirling as you'd see in a tornado, can easily exceed 60 miles per hour and lead to widespread tree and power line damage.

Downbursts are sudden rushes of air from the base of a thunderstorm that crash into the ground and radiate out like a ripple on a pond. A downburst usually forms when an updraft suddenly collapses, forcing rain and cold air to race toward the ground, or when dry air mixes into the downdraft, rapidly cooling the air and forcing it to fall to the ground at a high speed. Microbursts, which are strong downbursts that occur over a small area, pose a grave threat to airplanes as they're taking off and landing.

## Lightning

Lightning makes a storm's thunder. A budding cumulonimbus cloud is full of countless water droplets and ice crystals that bump into one another as they bounce around inside the clouds. All that bumping and scraping induces static electricity that leads to an excess amount of positive and negative charges in and around the clouds. Eventually, groups of charged particles build up to the point that they discharge the pent-up energy in a burst of heat that can reach temperatures of 50,000°F—hotter than the surface of the Sun. This momentary blast of heat turns the air to plasma, resulting in the visible flash. The heated air around a lightning bolt rapidly expands and creates a sonic boom, which we hear as thunder. The loudness of thunder depends on lots of

# STRUCK BY LIGHTNING WHILE INDOORS

It's true that you can be struck by lightning while you're indoors. The risk is relatively low in well-built homes that have plastic pipes and grounded electrical systems, but there are plenty of true stories about lightning entering homes through pipes, wiring, and windows. Landline telephones and fixtures that receive running water through metal pipes are the most dangerous places in a home when a thunderstorm creeps up. Lightning can strike the house or the ground nearby and easily zap someone talking on the phone or taking a bath.

Lightning striking people inside their homes is still a huge concern around the world, especially in countries where many homes are constructed with poor electrical wiring or bare dirt floors. This is true even in the United States, where nearly one-third of all lightning strike injuries occur while the victim is indoors. There are worldwide efforts to provide people in these regions with the weather equipment and tools necessary to keep safe when lightning approaches.

different variables, including the lightning strike's distance from you, the shape of the lightning bolt, the terrain in the area, how heavily it's raining, and the air temperature.

Storms produce billions of lightning flashes around the world every year. A single thunderstorm can produce tens of thousands of lightning strikes over the course of its life. We can see lightning strike within a single cloud (intracloud), lightning strike between clouds, and lightning strike from the cloud to the ground. Intracloud and cloud-to-cloud lightning usually lights up the sky with a brilliant flash. A long, clear view of the horizon makes cloud-to-cloud lightning a treat during a nighttime thunderstorm, allowing observers to watch intricate webs of lightning crawl across the sky.

Cloud-to-ground lightning is the type we have to worry about. Positive charges building up at the base of a thunderstorm cause negative charges to build up along the ground. Electricity follows the path of least resistance, so lightning typically strikes tall objects such as radio towers, buildings, and trees. However, the greatest buildup of negative charges can make a target of vehicles, people caught outside, or even the ground itself. Some lightning bolts can even travel many miles away from the storm, leading to the infamous "bolt from the blue" that can strike beneath clear skies.

Most of the world's thunderstorms occur in the tropics, since the soupy equatorial regions provide thunderstorms ample fuel to bubble up. Strong thunderstorms can even occur in northern Canada and interior Alaska at the peak of summer's warmth. However, a handful of areas struggle to see many thunderstorms at all. Much of California can go years without ever hearing a clap of thunder. The cold waters of the eastern Pacific Ocean keep the air relatively stable near the surface, limiting the opportunities for convection. It usually takes a strong low-pressure system to provide enough lift to trigger thunderstorms in California.

# TORNADOES

Tornadoes are the ultimate expression of the sky's unstoppable power. These mammoth whirlwinds can tear across the landscape with ease. It's ironic that such beautiful storms can wreak such complete destruction. Thousands of people flock to the Great Plains in the central United States every spring to see these windstorms in all their might, while thousands more in the path of those tornadoes take cover in basements and bathtubs and hope to see clear skies behind the storm.

A tornado is a rapidly rotating column of air that stretches from the base of a thunderstorm all the way to the ground. Most tornadoes begin and end their lives as funnel clouds, which are similar structures that don't make contact with the earth. While most tornadoes are funnel clouds at some point in their life cycle, not all funnel clouds turn into tornadoes. Some funnel clouds form and dissipate harmlessly without ever touching down.

The visible portion of a tornado is called the condensation funnel. This is the cloud that forms around the outer edge of a tornado as a result of the low air pressure within the twister itself. It's similar to the cloud that forms at the top of a soda bottle when you first open it on a humid day.

You can't always see tornadoes. The strong winds blowing around large tornadoes can extend a great distance from the condensation funnel, sometimes even catching storm chasers by surprise. Tornadoes that form in dry environments can complete their entire life cycle without ever developing a condensation funnel. The only hint that they're there is a rotating cloud base and swirling debris just above the ground.

Few tornadoes are exactly alike. The average twister is a few hundred yards wide, while the largest on record clocked in at more than two miles wide. Meteorologists and storm chasers have come up with different terms to describe their shapes and sizes. Rope tornadoes look like tendrils weaving and wiggling across the sky. Stove-pipe tornadoes look smooth and even from top to bottom. Wedge tornadoes, the largest variety, look like black walls rolling across the horizon, consuming entire neighborhoods at once.

Size doesn't always relate to wind speed. Even a tiny tornado can produce winds strong enough to send debris flying at lethal velocities. Since it's hard to measure the winds in a tornado directly, scientists developed the Enhanced Fujita Scale to estimate

# GREEN SKY BEFORE A TORNADO

One of the longest-lasting weather adages in American life is to look out for a green sky during a thunderstorm, as it's a telltale sign that a tornado is on the way. While it's not exactly accurate, it contains a kernel of truth.

The sky can feature a whole palette of threatening colors as a powerful thunderstorm approaches. There's an immense amount of water and hail suspended in a fresh thunderstorm. The contents of a thunderstorm determine how much sunlight gets filtered out. The thickest, juiciest storms block out enough light to leave the sky pitch black. A bluish or greenish sky can signal a thunderstorm that's rich in hail. It takes a strong updraft to keep heavy hailstones suspended high in the sky, and strong updrafts are needed for the development of tornadoes.

# THE LARGEST TORNADO

The largest tornado ever recorded touched down outside of El Reno, Oklahoma, on May 31, 2013. National Weather Service meteorologists who surveyed the damage found that the tornado reached an astonishing 2.6 miles wide. The tornado was so large that it appeared as a solid black mass trudging across the prairies of Oklahoma. The storm killed eight people, including several storm chasers who were overtaken by the tornado's rapid expansion.

A mobile weather radar near the tornado measured maximum winds of nearly 300 miles per hour, which would make this one of the strongest tornadoes ever recorded. However, the tornado is officially rated an EF-3 because the Enhanced Fujita Scale is based on damage, and meteorologists found no damage to justify a scale-topping rating. The El Reno tornado came just two weeks after an EF-5 tornado devastated the city of Moore, Oklahoma, just a few dozen miles to the east.

the maximum strength of a tornado based on the damage it leaves behind. The scale ranges from EF-0 to EF-5. The lowest rating covers broken tree branches and shredded roof shingles, and the highest rating is reserved for tornadoes that can rip a well-built home clean from its foundation. Since 1950, only a few dozen tornadoes on record have ever grown strong enough to reach the top of the scale.

Just about every country on Earth is prone to tornadoes, though they're most common in the middle latitudes where the weather swings through all four seasons. The United States sees the bulk of the world's twisters, recording more than 1,000 in a typical year. Canada, Argentina, Bangladesh, South Africa, Australia, and the countries of central Europe see a heightened risk for tornadoes during their stormy seasons. Bangladesh and northern Italy are prone to large and violent tornadoes like the ones seen in Oklahoma or Kansas.

The United States is exceptionally vulnerable to tornadoes due to its unique geography. The country lies at just the right latitude to see the jet stream spawn low-pressure systems on a regular basis. Warm, humid air flowing north from the Gulf of Mexico provides ample fuel for thunderstorms to explode when cold fronts and dry lines sweep in from the west. This atmospheric battleground typically sets up over the central plains—commonly known as Tornado Alley—and the Southeast, though tornadoes occur regularly in all 48 of the contiguous United States. Conditions that lead to tornado outbreaks don't only come around in the spring, either. Twisters are a year-round threat in the United States; some of the deadliest outbreaks in the southern states have occurred during the long dark of a winter's night.

Most of the tornadoes that touch down around the world form from supercell thunderstorms. Supercells, driven by rotating updrafts, are a remarkable sight when you witness one up close. A well-formed supercell almost looks like a spaceship hovering in the distance, with many layers of smooth, spinning clouds piled high on top of one another.

The rotation within a supercell is important in the development of tornadoes. A particularly strong supercell can see its updraft narrow and strengthen. A tighter updraft can begin to stretch down from the base of the thunderstorm and spin faster as a result of conservation of angular momentum, the same principle behind a figure skater spinning faster when they pull their arms in toward their body. The resulting funnel cloud continues descending toward the ground until it touches down as a tornado.

# TORNADO SAFETY MYTHS

In the days before we had modern radar technology and an understanding of how these destructive whirlwinds form, tornadoes were especially terrifying. The dearth of knowledge before the 1960s led to widespread misconceptions about tornado safety that have been repeated as fact through the generations.

One common fallacy is that someone in a tornado's path should open up all the windows to equalize the pressure between the outdoors and the indoors. While homes struck by tornadoes may appear to explode, it is caused by the force of the wind shredding them apart, not internal pressure. In fact, opening the windows makes it easier for wind to get inside and damage a structure.

Another myth is that folks under a tornado warning should head to the southwest corner of the basement. This advice arose from the fact that most tornadoes in the United States move from southwest to northeast due to the prevailing winds during classic severe weather outbreaks and the cyclonic nature of supercell thunderstorms. However, sheltering in the southwest part of the basement doesn't necessarily protect you from flying debris, which swirls in every direction. The main goal is to get deep underground to put as many walls between you and flying debris as possible.

A supercell's tornado can continue until the downdraft of cold air chokes off the updraft of warm air, disrupting the tornado and forcing the storm to weaken. Strong supercells can avoid choking themselves off for many hours at a time, allowing a tornado—or a family of several successive tornadoes—to track along a path hundreds of miles long.

Supercells aren't the only culprit behind tornado formation. Smaller, weaker tornadoes are common in the kinks that develop along the leading edge of squall lines or in the outer bands of landfalling tropical cyclones. These tornadoes, often called "spin-up" tornadoes in weather forecasts, can form and dissipate within a couple of minutes, giving forecasters little or no time to issue a warning before they strike.

Even large tornadoes are tough to predict. While weather models are advanced enough that forecasters can accurately predict where conditions are favorable for rotating thunderstorms and tornadoes, they still can't predict exactly where or when a tornado will touch down. They also still don't know why one supercell thunderstorm will produce a tornado while another identical supercell doesn't.

Even as scientists continue researching better ways to predict tornadoes before they form, incredible advances have been made in detecting the rotation within thunderstorms in the minutes before a tornado develops. The US National Weather Service upgraded the country's network of weather radars in the 1990s, using Doppler technology to calculate the speed of raindrops, snowflakes, and hailstones within a thunderstorm. This advance gave forecasters the ability to see rotating winds within a thunderstorm, leading to a dramatic leap in tornado warning lead time that's saved countless lives in the decades since. A recent technological upgrade known as dual-polarization now shows the size and the shape of objects detected by the radar beam, allowing forecasters to spot tornado debris swirling around inside a thunderstorm.

# WIND

The unmistakable roar of a formidable windstorm is unforgettable. The walls shudder and the roof creaks, trees bend and sway as air presses through the leaves with a bold *whoosh*. On most days, it's easy to forget that winds during a storm can be strong enough to knock us over. These air currents are the root of almost all of our weather, and the processes that move the air around us are just as versatile as the conditions they can create.

Wind is the product of imbalance. The breezes that mess up our hair are the result of air rushing around in an attempt to empty out areas of high pressure and fill in areas of low pressure. Once a center of low or high pressure finally equalizes, another excess or dearth of air develops somewhere else, and the wind blows once again.

In most cases, jet streams are ultimately responsible for the winds we feel at the surface of the earth. These strong belts of wind in the upper atmosphere generate the strong, sustained lift needed to power a center of low pressure. The jet stream is strongest during the late winter and early spring, when the subtropical latitudes begin warming up as everywhere else is still struggling to fight off the final throes of winter.

Since spring's warmth encroaching on winter's chill creates a steep temperature gradient that strengthens the jet stream, we wind up experiencing more intense low-pressure systems at that time. The pressure difference between the center of a budding storm system and the environment around it leads to winds that can roar with memorable fury.

## Wind Builds, Land Brakes

A few times a year, winds blow hard enough that venturing outside seems intimidating, especially over the Great Plains. The actors in the musical *Oklahoma!* don't sing about

# HEAT BURST

Dying thunderstorms are usually a welcome sight for folks who've had enough interesting weather for one day. However, these storms sometimes have a rare but nasty trick up their sleeves. If a pocket of dry air intrudes into the middle of a thunderstorm, the dryness can evaporate the raindrops and leave cooler air behind in their wake. This cooler, less-dense air sinks through the thunderstorm and picks up speed as it evaporates more raindrops in its path. The sinking air warms up as it compresses and races toward the ground.

When the falling air hits the ground and spreads out, the sudden rush of hot wind, called a heat burst, can send nighttime temperatures soaring above 90°F. A collapsing thunderstorm over Wichita, Kansas, sent the temperature soaring to 102°F just after midnight on June 9, 2011. While meteorologists are still studying why some thunderstorms generate heat bursts while others don't, most of the notable instances in the United States were recorded in the central part of the country during the nighttime hours.

the wind sweeping down the plain for nothing. The land in the center of the United States is flat. Extremely flat. The level terrain and lack of trees means that you can see for miles and miles, as far as your eyesight and the stalks of corn and wheat will allow. All this flat, relatively featureless land is a perfect environment for wind to flourish.

Land slows down wind through friction. All the trees, grasses, buildings, and other objects that dot the landscape act like little sails that catch the wind and slow it down. Friction forces winds to blow considerably slower at the surface than they do a few thousand feet up where there's hardly any influence from the ground itself. You can see this in action when you feel a soft breeze but low-hanging clouds speed along above you.

The wind that's zipping above our heads doesn't always stay up high. Warm sunshine heats up air near the ground and causes the lowest layer of the atmosphere to churn, producing vertical air currents that can shove stronger winds from higher levels down to the surface  and make a windy day even windier. It's not uncommon for wind gusts to measure twice as high as sustained winds. Instability wanes after sunset when the air cools, which is why winds tend to calm down at night.

## Mountain Winds

Mountains are buffeted by much stronger winds than places at lower elevations, where friction with the land itself helps slow the wind down. Higher up, though, there's little to slow winds down before they slam into the side of tall mountains.

Scientists record weather observations at a station at the summit of New Hampshire's Mount Washington, which is the tallest mountain in the northeastern United States. The mountain's peak, which juts nearly 6,300 feet above sea level, experiences some of the worst weather conditions on Earth. Gusts atop Mount Washington can exceed 100 miles per hour as vigorous storm systems move over New England. The greatest wind gust recorded here reached a record-setting speed of 231 miles per hour during a storm on April 12, 1934.

It's not just mountain peaks that take the brunt of winds as they twist and bend around the high terrain. Winds that blow over a mountain have to descend the other side, and that can cause plenty of problems for folks who live downwind of a tall range such as those in eastern Colorado and densely populated areas of Southern California. A strong downsloping wind on the leeward (downwind) side of the Rocky Mountains

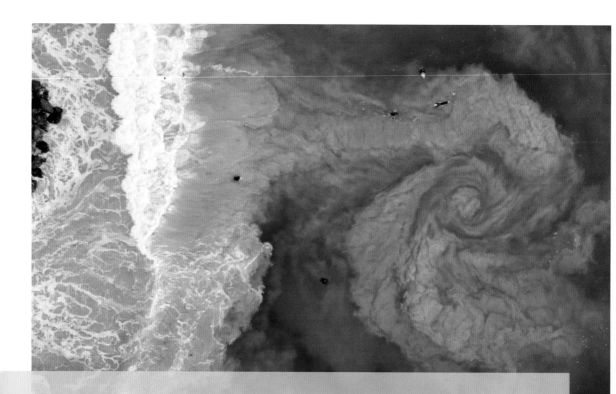

# RIP CURRENTS

Folks heading to the beach to cool off on a warm day sometimes hear warnings of rip currents that could pose a threat to their safety. This strong flow pulls water out away from the beach and can take swimmers with it.

Rip currents are generated by winds that blow waves head-on into the beach. Waves that hit the beach at an angle can drain their water out to sea without much fuss. When a wave breaks straight into the beach, though, the runoff has nowhere to go but straight out into the ocean. The area between two such waves can turn into a powerful current that moves faster than the fastest swimmer can swim.

These strong flows don't pull you underwater like they show in the movies, but they are dangerous because swimmers can grow exhausted when they panic and try to swim against them. The best way to escape a rip current is to swim parallel to the shore until the water stops pulling you away from land, then cautiously swim back to safety. The shape of the coastline and prevailing winds make some beaches more prone to rip currents than others. Playa Zipolite in southern Mexico, Hanakapiai Beach on the Hawaiian island of Kauai, and numerous beaches around Australia are exceptionally vulnerable to strong rip currents.

is referred to as a Chinook wind. A Chinook event occurs when the winds of an approaching low-pressure system flow over the mountains and race down the other side. Air pressure increases as the air descends from the top of a mountain to the bottom. That's why we have to pop our ears to relieve the pressure when we drive down a mountain. The pressure exerted on descending air causes that air to heat up and dry out as it flows toward lower elevations. This hot wind—a Chinook—often results in clear, warm, and blustery conditions for communities like Calgary, Alberta, and Denver, Colorado.

Santa Ana winds are a similar regional wind event, which plays out in Southern California. Clockwise winds wrapping around a high-pressure system over the deserts of Nevada and Utah send wind up and over the mountains east of Los Angeles. These hot and dry winds are ferocious by the time they reach the coast, sometimes sending temperatures soaring above 90°F and bringing bone-dry humidity levels that can dip into the single digits. Conditions like these are ideal for fast-spreading wildfires that can quickly consume thousands of acres of land.

## Winds Glide over Water

Wind thrives over water. Air encounters low friction when it blows over ponds, lakes, and oceans. The near-constant flow of air over bodies of water has shaped the course of human technology and exploration. Even today, the standard unit of measurement for wind over water is knots, rooted in the days when ships measured their speed by unspooling a knotted rope behind the moving vessel and counting how long it took for the equally spaced knots to pass over the end of the ship as the rope unfurled.

There's a fine line between just enough wind to sail and too much wind to stay afloat. Strong windstorms over the oceans, often called gales, can churn the surface of the sea so much that giant waves can overtake and even capsize ships. Stronger winds and a longer fetch—distance traveled over open water—can create stronger waves. Several cruise ships in recent years found themselves at the mercy of large storms that generated waves so large that the vessels rocked like toys in a bathtub. Wind-driven waves can also pose a serious hazard to coastlines, leading to devastating beach erosion and rip currents.

# TEHUANTEPECER

Raging winds blowing south over the Gulf of Mexico eventually run out of water to roam free. Air rushing southward runs up against the mountains of southern Mexico. There's a narrow gap in the mountains called Chivela Pass that allows air to squeeze through as it blows over the isthmus of Tehuantepec, a narrow strip of land that separates the Gulf of Mexico from the eastern Pacific Ocean.

Winds speed up as they press through this tiny strait, leading to a powerful Pacific wind called the Tehuantepecer. This wind is ferocious, blowing at more than 50 miles per hour during strong events. The Tehuantepecer is so strong that it can pose a threat to boats in the region, and the waves can stir the ocean and cool off the warm Pacific waters below.

## Sea Breezes

Coastlines witness one of the most fascinating interactions between the land, water, and sky. During the warm months, sea breezes cycle daily between blowing ashore during the day and blowing back out to sea at night.

Sea breezes are a product of heating. Land heats up during the day much faster than the water, causing air over the ground to rise through convection. Rising air leaves lower air pressure over land, forcing cooler air over the ocean to rush inland to equalize the pressure imbalance. This onshore wind is a daily occurrence in coastal areas, and it's a major source of the regular thunderstorms that drench seaside communities during the spring and summer.

Florida's peninsula experiences two sea breezes during a typical day: one from the Gulf of Mexico to the west, and one from the Atlantic Ocean to the east. These two sea breezes typically converge near the middle of the state, leading to intense daily thunderstorms over cities like Orlando.

This process happens in reverse overnight. Land cools off quickly after sunset while water temperatures stay the same, allowing air temperatures over the water to stay warmer than the air over the land. This warm ocean air rises and causes wind to blow from the land out to sea, resulting in a land breeze that can trigger thunderstorms dozens of miles offshore. If you ever visit a warm coastal destination and wake up around sunrise, you'll probably be able to spot flashes of lightning on the horizon when you look out over the ocean.

# SUMMER

Summer is the slowest of the four seasons weather-wise. Aside from the occasional hurricane roaring ashore, the big story during those sultry months is usually the lack of any meaningful weather. Deserts broil under the hottest temperatures possible in the natural world, while cities in temperate areas stew in choking humidity that's hard to escape. Even though summer days aren't usually dominated by destructive storms, the low simmer of relentless heat and humidity can take a toll.

# SUMMER SOLSTICE

**S**ummer is all about the heat. There are days that make even the hardiest warm-weather person thankful for the invention of air conditioning. It's the one time of year when a distinct lack of weather often leads to the harshest conditions. Folks in the Northern Hemisphere mark the beginning of summer near the end of June when the Sun's direct rays shine down on 23.5°N latitude. This line, the Tropic of Cancer, lies not far from the southern tip of Florida. While the hot sunshine looms the highest in the sky on that date, it takes a while for the heat to build into an unbearable blanket.

Winter's chill can settle in long before the date of the winter solstice, but the toastiest summertime temperatures often don't reach their peak until the middle or even the end of the season. Summer's heat becomes unrelenting in July and August in the Northern Hemisphere because the jet stream, which thrives on strong temperature gradients, retreats north during this time of year.

Lower latitudes gradually warm up through the beginning of summer, which in turn gradually pushes the jet stream farther north. There's not a very big temperature gradient between the south and the north once we're in the heart of July, so the jet stream is forced all the way into Canada. This leaves only communities in the northern United States and southern Canada in the path of storm systems that stroll along in the westerly flow.

# Everybody talks about the weather, but nobody does anything about it.

—CHARLES DUDLEY WARNER,
*Hartford Courant*

Without the bold influence of the jet stream, summertime weather is dominated by smaller-scale features like thunderstorm complexes and hurricanes. Occasionally, a large upper-level ridge of high pressure will form. These ridges are sometimes nicknamed a "ring of fire" because of the brutal heat and stifling humidity that can build beneath them.

A spell of high heat and humidity that lasts several days is called a heat wave. Much like snowstorms, different regions have different definitions of heat waves, depending on what locals are used to handling. Heat waves in Vermont feature lower temperatures than heat waves in Arkansas. But it's not just the heat that makes a heat wave so tough to handle—it's the humidity.

## Relatively Bad Humidity

Humidity describes the amount of moisture in the air. It's an important part of weather forecasting, especially during the summer, when high humidity can make a day feel flat-out miserable.

Relative humidity is a widely used measurement of how much moisture is present in the air based on the air temperature. This metric always shows up in weather reports, and it has secured its spot as a mainstay of small talk. It's a sure bet that you can walk into a grocery store during the summer in a place like New Orleans and hear a sweaty shopper exclaim that it's 90°F out there *with 60 percent humidity.* Mercy!

The widespread use of relative humidity is enough to frustrate any meteorologist. This ubiquitous percentage is a bad metric to use to talk about how moist or dry the air feels. We call it "relative humidity" because it tells us how much moisture is in the air relative to the air temperature. The air reaches 100 percent relative humidity, or full saturation, when it can't hold any more water vapor than what's already present.

Since air condenses when it's cool and expands when it's warm, there's less room between air molecules for water vapor to occupy on a cold morning than on a hot afternoon. As a result, relative humidity goes up in the morning and falls through the afternoon as the day heats up. Going solely by relative humidity, it's always going to seem more humid in the morning than it does in the afternoon, even if the air feels like a sauna under the scorching sun.

Relative humidity is useful for two purposes: wildfire forecasts, where the ignition point of vegetation depends on specific moisture levels, and fog forecasts, as air that is fully saturated is likely to form fog. That's about it.

## Embrace the Dew Point

The best way to talk about the moisture in the air is to use the dew point, which is the temperature air has to cool to in order to condense and create dew. A 50°F dew point is the same whether the air is refreshingly cool or scorching hot. This consistency makes the dew point the best metric to gauge how dry or moist the air feels without having to do advanced calculations in your head.

Lower dew points indicate more comfortable humidity, while higher dew points occur on those muggy days that hit you like a wrecking ball when you step outside. As a general rule of thumb, dew points lower than 50°F indicate dry air. It's noticeably humid when the dew point climbs above 60°F. The air starts to feel uncomfortably sticky when the dew point climbs above 65°F, and it's downright tropical when the dew point reaches 70°F or higher. It's possible to see dew points exceed 80°F near moist corn fields on an exceptionally hot day or in desert areas that lie near bodies of water such as the Persian Gulf.

## Break a Sweat

"It's not the heat that gets you, it's the humidity!" While it's best to politely nod when Aunt Linda whips out that tried-and-true summer conversation starter, it's the rare cliché that's backed up by science. A brutally hot and muggy day is much tougher on the human body than a day that's just plain hot.

The human body cools itself off through sweat. Water absorbs heat when it evaporates. Our bodies take advantage of this physical principle by releasing sweat that evaporates into the air and cools the skin below, which serves to lower our body temperatures to a manageable level.

Humidity messes with this process. When the dew point is high, it's tough for the air to hold more moisture than it already contains. The excess mugginess means that sweat can't evaporate efficiently, so it just sits on our skin and makes us feel all sticky and clammy. Pooling sweat does more than make us feel gross. Without the ability to cool down, our body temperatures can climb dangerously high if we're outside too long on a steamy day.

# SUMMER'S SOUPY POLLUTION

For most of the twentieth century, summer in the United States was synonymous with choking air quality on those stagnant, sultry days. Concentrations of pollutants like vehicle exhaust, smoke, and factory emissions can build up to dangerous levels on a hot day when winds are light and humidity is high.

The Air Quality Index uses a color-coded scale to measure the potential health risk of the air quality on a given day. A "code red air quality day," for instance, indicates there are enough pollutants present that spending a long time outside could cause serious problems for folks with respiratory conditions, and even threaten those who are perfectly healthy.

The number of days at code red and code purple (the highest category on the scale) across the United States has steadily declined in recent summers as stricter environmental regulations and greener technology have reduced the number of harmful emissions spilled into the air by vehicles and industries.

Anyone who's out in the heat—whether it's humid or dry—for too long without cooling off runs the risk of serious injury from heat-related illnesses. Heat exhaustion sets in when your body begins to overheat and your system kicks into overdrive to try to cool you down. Muscle cramps, a rapid pulse, excessive sweating, and weakness are signs that it's time to drink water and get to shade right away. Heat exhaustion can give way to heat stroke, a life-threatening emergency that arises when your body's essential functions begin shutting down because of high body temperature and dehydration.

> What dreadful hot weather we have! It keeps one in a continual state of inelegance.
>
> —JANE AUSTEN, in a letter to her sister, 1796

## Feel the Heat Index

Extended heat waves are dangerous because water is really good at retaining heat, which is why muggy days don't cool off much at night. The lack of relief after sunset compounds the effects of heat and humidity on the human body to tragic effect. Day after day of stifling temperatures followed by an uncomfortably warm and humid night quickly takes a toll on individuals whose bodies can't take the stress.

Big weather events like tornadoes can cause destruction and death, but even just hot weather can be dangerous. Heat is one of the leading weather-related causes of death in the United States every year. Strong and prolonged heat waves are exceptionally dangerous to vulnerable populations such as the elderly, low-income families who can't afford fans or air conditioning, and folks who suffer from chronic illnesses.

The infamous Chicago heat wave of 1995 killed more than 700 people in just a couple of days, taking a horrendous toll on the city's low-income and elderly populations. Heat waves in Europe are exceptionally deadly because summers are normally temperate across most of the continent, precluding the need for widespread air conditioning.

Recognizing the need to convey the seriousness of the pairing of heat and humidity, meteorologists worked with medical experts to develop the heat index. This combines the actual air temperature and the dew point to tell you what the combined effects of high heat and high humidity feel like to your body.

If it's 100°F out with a dew point of 72°F, the heat index would be 111°F. Even though the actual air temperature is cooler, the high humidity makes your body work as hard to cool you off as it would if the air temperature were 111°F. Heat-related illnesses can set in faster the higher the heat index climbs.

While the heat index is standardized and can be used anywhere in the world, different countries may use their own variations of the heat index to warn residents about potential harm. Environment and Climate Change Canada, the country's governmental weather agency, uses its own metric called the "Humidex" to relay how warm it feels when you factor in the humidity.

# THE YEAR WITHOUT A SUMMER

Mount Tambora may be a small mountain on a small island in Indonesia, but the volcano is responsible for the "Year Without a Summer" in much of the Northern Hemisphere. The volcano blew its stack in 1815, one of the strongest eruptions in recorded history. The fallout from the explosion killed tens of thousands of people who lived around the volcano, but the danger didn't end with streams of lava and landslides of hot ash.

The eruption of Mount Tambora gushed so much debris into the air that a thick layer of ash remained suspended in the atmosphere for more than a year. The ash reflected enough sunlight back into space that it caused a decline in global temperatures that temporarily altered weather patterns. The effect was so pronounced that, in 1816, parts of New England saw snow in June and temperatures cold enough to see your breath through the rest of the summer. Conditions slowly returned to normal as ash finally succumbed to gravity and the skies cleared out.

# HAIL

Hail is the oddball of the early summer skies. Rarely seen during the winter months, this frozen precipitation falling from thunderstorms can range from the size of a pea to a chunk of ice so large that it leaves a divot. When it falls on a hot day, it feels as if the hail's existence violates one or two of the laws of physics. However, it makes perfect sense when you consider what's going on inside of a cumulonimbus cloud.

# WHAT'S IN HAIL?

After every big tornado comes a slew of stories about folks hundreds of miles away from the destruction finding personal items like paperwork and photographs wedged in a bush outside their front door. The intense updrafts in and above tornadoes allow these fearsome storms to lift debris and toss it incredibly far distances.

Objects lifted into the heart of a tornadic thunderstorm can serve as the nucleus for hail formation. It's not terribly uncommon after a big tornado to find hailstones that formed around tree branches, splinters of plywood, and other pieces of debris sucked into the heart of a thunderstorm.

## How Hail Forms

Thunderstorms thrive when there's a big temperature difference between air at the surface and the air aloft. Within a budding thunderstorm, temperatures often dip below zero near the top of the storm. Ice crystals begin to form within these frigid temperatures. When liquid cloud droplets condense onto an ice crystal in this subfreezing air, the newly formed droplet freezes into the core of a hailstone.

These hailstones get caught in the thunderstorm's strong updraft inside a cumulonimbus cloud. Rapidly rising air tosses them to the top of the storm and they begin to fall, only to get caught by the updraft again and go for another ride through the storm. As they rise and fall through the cloud, they're coated in liquid droplets that freeze as a new layer of ice. Hailstones grow one thin layer of ice at a time for many minutes as they bounce around inside of a thunderstorm. Hail falls to the ground when it wanders too far away from the updraft to stay aloft, or when the hailstone itself grows too heavy for the updraft to support its weight any longer.

## Hail Size Matters

Meteorologists prefer to measure hail precisely, but it's often hard to find a caliper while tornado sirens are blaring, so they use common object comparisons. Most hail is tiny and either melts before it reaches the ground or makes it to the surface without causing any damage. Pea-size hailstones land with a "tink" and provide some entertainment in yet another summer torrent.

Hail begins to cause damage when it reaches one inch in diameter, or about the size of a quarter. That's the point at which hail can leave dents on cars, damage roofing materials, and possibly even break windows.

The potential for damage increases with hail size. Golf-ball-size hail is relatively common in the United States. These iceballs easily fill the palm of your hand and can leave so much damage to vehicles that insurance adjusters will declare them irreparable.

Hail is exceptionally dangerous when it grows any larger than a golf ball. Supercell thunderstorms on the Great Plains can produce hailstones the size of baseballs or even larger. These hailstones can decimate crops, total thousands of vehicles, and even punch holes in the walls and roofs of homes and businesses.

Between 1955 and 2019, the US Storm Prediction Center received 217,364 reports of hail one inch in diameter or larger. Nearly 40 percent of all those reports came in for hail the size of golf balls or larger, with 5 percent measuring the size of a baseball or larger. Just under 1 percent were at least four inches in diameter. The largest hailstone ever recorded fell in Vivian, South Dakota, on July 23, 2010. The record-breaking chunk of ice measured eight inches in diameter and weighed nearly two pounds.

It's not just the size of hail that matters when it hits the ground—it's how fast it's moving by the time it gets there. Hail falling straight down out of a thunderstorm can reach the surface traveling at highway speeds. If you add in the powerful winds that severe thunderstorms often generate, a hailstone can reach its destination traveling more than 100 miles per hour. Hailstones blown by the wind can easily shatter windows, which is why forecasters warn listeners to stay away from windows and exterior walls during a severe thunderstorm.

The shape of a hailstone varies as much as its size. Whether a hailstone is smooth or spiky, perfectly round or pleasantly oblong depends on different environmental factors within the storm itself. Hailstones that are wet can collide and freeze into larger conglomerate hailstones that break apart when they hit the ground. These collisions are responsible for the abstract, jagged appearance of some larger hailstones.

The speed at which water freezes on a hailstone can even affect its appearance. Hail turns milky and opaque when water freezes quickly and traps lots of tiny air bubbles within the ice. If the water freezes slowly enough that the air bubbles escape, it leaves behind ice clear enough to use in a cocktail.

## Where Hail Forms

Any type of thunderstorm can produce hail; even a pop-up thunderstorm on a hot summer day can lay down a quick blanket of hail if there's enough cold air aloft. Hail, especially the big kind, is most common within a supercell thunderstorm. The intense rotating updraft that drives a supercell can support the weight of large hailstones, which is why the threat for significant hail follows the strongest thunderstorms around the world.

Major hailstorms are common in the United States, parts of central and eastern Europe, South Asia, eastern Australia, southern Africa, and the pampas of South America. All of these regions experience weather patterns that make supercells possible, carrying with them the potential for hail (as well as tornadoes).

Across the United States, hail begins to pose a serious problem in the southern states early in the spring, as rounds of severe thunderstorms traverse the land from

## DEATH BY HAIL

Hailstorms have a remarkably low mortality rate compared to most types of severe weather. Better technology, better construction standards, and even the gradual migration of most jobs indoors have cut down on the opportunity for hail to cause serious injury or even death. But that's not always enough to keep folks safe.

Hail kills by striking a fatal blow to the head. Most stones don't fall with enough force to cause instant death, but any head wound is serious, especially for someone with pre-existing medical conditions.

There's a gruesome history of hail taking a serious toll when folks are caught in the wrong place at the wrong time.

Hailstorms in China, Bangladesh, and India have killed dozens of people in recent decades. The last mass-casualty incident in the United States occurred in Dallas, Texas, on May 5, 1995. A severe storm crept up on an outdoor festival with thousands of people in attendance. Not only was it one of the costliest hailstorms in American history, but the storm injured hundreds of people and hurt 60 people seriously enough that they had to be hospitalized.

Texas to the Carolinas. As the season warms up and turns to summer, the greatest extent of severe weather moves into the traditional "Tornado Alley" states on the central and southern plains. Storms that can support huge hail are endemic to the northern plains states during the height of the summer, where there's still enough wind shear and cold air aloft to support supercells.

## Spotting Hailstorms

Finding yourself outside during a thunderstorm is dangerous no matter the circumstances, but getting stuck outdoors when hail begins is a life-threatening ordeal. In the days before weather warnings and Doppler radar, people were often caught by surprise by hailstorms, often leading to injuries or worse. Today, it's pretty rare to hear about serious injuries inflicted by falling hail, a testament to improved forecasts and warnings in recent decades.

Modern weather radars are capable of spotting hail in a thunderstorm even before it begins to fall. Radars in the United States sweep a beam of energy through the sky at different altitudes every couple of minutes. Computer programs analyze these layers of radar data to compile vertical cross sections and 3D plots of the atmosphere, which can allow a meteorologist to spot a growing hail core inside a thunderstorm.

The radar beam itself is designed to detect hailstones—and anything else floating around in the sky. During the 2010s, the National Weather Service retrofitted their weather radars to utilize dual-polarization technology. Before this upgrade, radars sent out a single beam of microwave energy to detect precipitation. Dual-polarization, as its name suggests, added a second beam of radiation to the radar dish. The new radar beam looks like a plus sign, which allows the radar to discern the size and shape of the objects in the beam's path. Since hail is much larger than individual raindrops and oddly shaped, it stands out on radar imagery, lending a helping hand to meteorologists on the hunt for hail within a thunderstorm.

# DESERTS
## and TROPICS

Through all of the boisterous storms and wild temperature swings, it's sometimes easy to forget that most weather ultimately averages out to a comfortable middle ground. Modern conveniences allow us to ride out a few weeks of brutal heat or bone-chilling cold. There are parts of the world, though, where the climate is so harsh that it outmatches even the best technology.

A location's climate is a product of its position on the planet and the geography that makes up the area. Average weather conditions steadily grow colder as you venture farther away from the equator. People in communities by the ocean usually experience more rainy days than folks who live thousands of miles from the nearest body of water. Two different towns separated by a mountain range can seem like they're a world apart when it comes to daily conditions. And regional climates are also determined by their position relative to the three major vertical air currents that circulate air from lower latitudes to higher latitudes in both the Northern Hemisphere and Southern Hemisphere. These air currents, which are responsible for creating the jet stream and can cover one-third of an entire hemisphere, are the Hadley cell, mid-latitude cell, and polar cell. All three transport air from one region to the next in the atmosphere's endless quest to achieve balance. The Hadley cell rises near the equator and sends air into the upper atmosphere, where it travels horizontally toward 30°N and 30°S before the air sinks back toward the ground. The mid-latitude cell rises near 60° latitude, or near the Arctic and Antarctic Circles, sending air up toward the poles and down toward the Hadley cell. The polar cells extend between 60° and 70° latitude to the poles and flow toward the lower latitudes.

The influence of these vertical air currents on regional climates is easily visible in pictures of Earth from space. Rising air over equatorial regions fosters a hot and steamy climate that routinely comes alive with vigorous thunderstorms. Since air warms up and dries out as it sinks through the atmosphere, regions located around 30°N and 30°S are frequently hot and arid, struggling to come across a drop of water to break the relentless heat.

## Deserts

A desert is an arid, barren stretch of land that's inhospitable to most life on Earth. Most imagine it as a sandy, dune-covered landscape scorching under the bright sunshine; deserts like that do exist in the Sahara in northern Africa and the deserts of Saudi Arabia. But not all deserts are hot. Many of the polar regions in northern Canada and Russia are considered deserts even though they experience some of the coldest air possible in the natural world.

True deserts receive less than 10 inches of rain in an average year. The lack of plentiful water over hundreds and thousands of years desiccates the ground, leaving

# Hadley Cell

The large-scale air currents that circulate over both the Northern and Southern
hemispheres have wide-ranging effects
on our climate. They generate jet streams,
foster jungles, and create vast deserts.

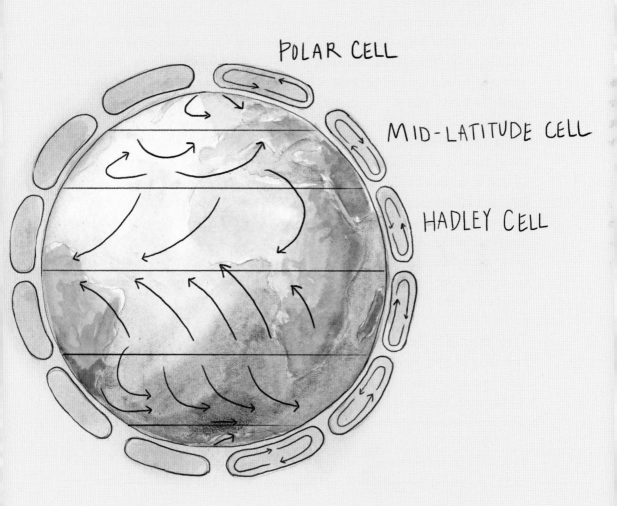

POLAR CELL

MID-LATITUDE CELL

HADLEY CELL

behind a cracked, cement-like landscape. Some plants and animals evolved to thrive in harsh desert environments; succulent plants like cactuses are thick and hearty, retaining what little water they can in order to make it through the long, harsh summers.

Looking at a physical map of Earth, you'll see that most of the world's deserts are all located around where air descends out of Hadley cells. The American Southwest, the Sahara desert, and the vast deserts of Southwest Asia all reside around 30°N, while South America's Atacama desert, the Kalahari of southern Africa, and Australia's Outback are all near 30°S. Closer to the North Pole, desolate tundra makes up the scenery around and north of 60°N.

While all deserts aren't hot, they can routinely be the hottest places on Earth. Descending air, intense ridges of high pressure, and barren land that reflects solar radiation all work together to drive up temperatures to unbearable levels. The world's hottest confirmed temperature peaked in a blazingly hot part of the California desert known as Death Valley. The observing station at Furnace Creek recorded a high temperature of 134°F on July 10, 1913.

The exceptionally dry air of a desert makes a hot day a little more bearable. Dry air allows sweat to evaporate from the skin efficiently, allowing the human body to cool off during quick trips outside. Nights in the desert always aren't too bad. Humid climates stay sticky and warm through the night because of water's high heat capacity. Remove the moisture from the air and it would get quite cool at night. The clear skies that let deserts grow so hot during the day are the same clear skies that radiate heat back away from the surface at night. It's not uncommon for some desert areas early in the summer to see a high temperature above 100°F during the day and a pleasant low temperature near 70°F at night.

## The World's Deserts

Each desert has its own character. The American Southwest is a bustling cultural and economic hub in the United States. The growing populations of metropolitan areas like Las Vegas, Nevada, and Phoenix, Arizona—each home to millions of people—are so large now that they're taxing the region's scarce water resources to the limit.

Australia's Outback is one of the largest and hottest deserts in the world. The continent's iconic red deserts get their rust-colored tint from centuries of iron-rich rocks weathering into sand that oxidizes when exposed to the air. The aridity and extreme heat in the center of Australia led to most of the country's population settling near the coastline.

Temperatures in Australia routinely climb above 110°F in the summer. The continent has been particularly hard-hit by climate change, enduring multiple extreme heat and wildfire events during the 2010s. The hottest day on record across Australia came during a heat wave in December 2019. The average temperature across the entire country on December 17, 2019, came to a blistering 107.4°F. This brutal reading broke the previous record of 105.6°F set just one day earlier.

Western South America's Atacama desert is so barren that it looks otherworldly. Space agencies send astronauts and scientists to the Atacama desert to simulate conditions on the Moon and Mars to study for future missions. The heart of the desert sits along the coast of northern Chile, stretching into portions of Peru and Bolivia. Northern Chile is the driest place on Earth outside of Antarctica. Some locations in far northern Chile receive less than 0.10 inches of rain in an average year, which amounts to little more than a fleeting sprinkle or light drizzle.

The Atacama is uniquely dry among the world's deserts due to a confluence of features that keeps the region rain-free. Not only does the desert lie beneath the descending air of a Hadley cell, but it experiences an intense rain shadow effect from the Andes Mountains to the east, blocking storm systems from providing the region much meaningful precipitation. The waters of the eastern Pacific Ocean experience upwelling—cold water from the depths of the ocean rising to the surface—along the Chilean coast, which keeps the air cold and stable over the Atacama.

One problem with keeping tabs on conditions in deserts is that there aren't many weather stations in such remote parts of the world. It's possible that we often miss global record temperature extremes because of the lack of reliable data collection in the deserts of Arabia or smack in the middle of the tundra. Meteorologists fill in these gaps using modern satellite technology to measure air temperatures. It's not as reliable as a standard observation with a thermometer perched just above the ground, but it's a decent way to track blistering extremes.

## Tropics

The same circulation that forces land into barren desolation also fosters lush rain forests that are so friendly to life that scientists are still discovering previously unknown species of flora and fauna. Regions around the equator are subjected to strong sunlight through-out the entire year, keeping equatorial areas consistently hot no matter the month.

Lush vegetation blankets just about every stretch of land near the equator from the dense rain forests of Central America to the jungles of Indonesia. The iconic rain forest of the Amazon River basin in northern South America is renowned for its size, density, and diversity.

The Amazonian rain forest's thick vegetation plays a significant role in the atmosphere itself, filtering the air of carbon dioxide and replacing it with oxygen. The survival of this rain forest—and regions like it around the world—is under significant threat as regional governments allow companies and citizens to raze and burn the vegetation to make room for construction and agricultural uses.

Daily weather conditions across the tropics are predictable. Intense sunshine allows cumulus clouds to dot the sky by late morning, evolving into heavy showers and thunder-storms by the middle of the afternoon. Thunderstorms wane with the heat of the day, and skies clear out in time to repeat the process by the following morning.

Conditions in the Amazon are hot and humid during the day and at night, with temperatures rarely varying more than a few degrees in either direction from one day or one season to the next. The heat and humidity exist in a feedback cycle that lends tropical climates their signature balmy air. Hot temperatures and bright sunshine induce plants into releasing their moisture into the air, which makes the air more humid and prevents cooling off at night.

Since tropical and desert regions don't experience defined seasons when it comes to average temperatures, these regions break down the year between dry seasons and wet seasons. While temperatures don't change, regional weather patterns respond to the changing seasons and bring distinct dry spells and rainy periods to tropical regions like India and arid regions like the American Southwest.

Vibrant expanses of greenery in tropical regions are fed by precipitation that forms within the Intertropical Convergence Zone (ITCZ). This is the region near the equator where trade winds from the Northern Hemisphere and the Southern Hemisphere converge with each other, triggering near-daily showers and thunderstorms. The ITCZ moves around throughout the year, bringing tropical seasons distinct rainy and dry seasons.

# ATMOSPHERIC RIVERS

Part of Earth's nonstop effort to strike balance in the atmosphere forces the skies to transport massive amounts of moisture away from the tropics and disperse it across drier climates. These thick ribbons of moisture, known as atmospheric rivers, snake around the world at high altitudes.

Acting like a reservoir in the sky, atmospheric rivers are responsible for exceptionally heavy rainfall. Thunderstorms can tap into these remarkable streams of water vapor and produce flooding rains in a hurry.

One of the most striking examples of an atmospheric river unfolds over the Amazon. The amount of moisture given off by vegetation in rain forests is immense. Brazil's National Institute of Amazonian Research estimates that plants in the Amazonian rain forest release billions of tons of water vapor into the atmosphere every day. Daily thunderstorms over the Amazon wring out some of that water vapor, while upper-level winds spread much of the moisture around the world. The next time it pours, you could very well see raindrops that originated from moisture deep within the Amazon.

# KÖPPEN CLASSIFICATION SYSTEM

The Köppen classification system isn't perfect, but it does a good job roughly outlining the many different climate areas across the world. Regions are broken down by groups and subgroups depending on latitude, elevation, annual temperature variations, and annual precipitation patterns. The six major groups cover tropical areas (Group A), desert areas (B), moist climates with mild winters (C), moist climates with harsh winters (D), polar climates (E), and highland climates (F) where the region's elevation—such as the Andes Mountains of South America and the Alps in Europe—

dominate annual weather patterns more than anything else.

Each of these groups can be divided into subgroups, producing more than a dozen different classifications that cover every climate you can encounter on Earth. Washington, DC, resides within Köppen classification Cfa, which means the city experiences a moist climate, mild winters, receives steady rainfall all year, and trudges through long and hot summers. San Francisco, California, sits within the Csb classification, which means they enjoy a moist climate, mild winters, and dry but cool summers.

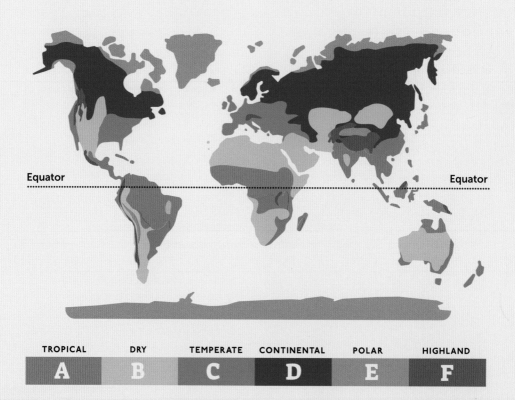

| TROPICAL | DRY | TEMPERATE | CONTINENTAL | POLAR | HIGHLAND |
|----------|-----|-----------|-------------|-------|----------|
| A | B | C | D | E | F |

## Monsoons

A monsoon is a seasonal shift in prevailing winds. A true-to-the-term monsoon occurs in India, which experiences a prominent wet season between June and September. Some towns near the mountains in northern India can measure many feet of rain during a single monsoon season.

Deserts can, surprisingly, have a monsoon season as well. The American Southwest experiences one during the summer months, as a high-pressure system drags humid air over the region from the Gulf of Mexico to the east and the Gulf of California to the south. The monsoon in the American Southwest isn't a monsoon in the original sense of the term, but the definition has expanded over the years to include the rainy season in desert areas.

Intense downpours over desert regions can lead to dangerous flash flooding. Unlike the rich soil you'd find in a more temperate area, the desert floor is hard and impermeable. Heavy rain doesn't easily soak into the dry ground, so even a small downpour runs off and pools up. Flash floods are a life-threatening hazard during the rainy season. Dry river beds, called arroyos, are particularly dangerous because they're attractive hiking grounds in places like Arizona and Utah. The runoff from a thunderstorm over the desert can flow many miles down these arroyos, bringing debris-filled floodwaters to areas where the skies are clear and sunny.

# DUST STORMS

**S**and gets everywhere. After a visit to the beach, it follows you home, hiding in every nook and cranny of your belongings for days. Now imagine it flying through the air. Blowing dust and dust storms are a serious problem for folks who live in dry regions of the world, posing a threat to the health and safety to those who find themselves in the wrong place at the wrong time.

The ground beneath our feet is the product of Earth's long history. Fertile soil is made up of minerals and centuries of plants and animals living and dying and replenishing the land that nourished them. Sand from desert areas is the product of rocks and minerals that have been weathered by years of wind, rain, and even the retreat of once-mighty oceans that covered the land.

Rich, healthy soil is held together by the root systems of grasses, brush, and trees that cover the landscape. Fertile soil has the structure and the moisture to withstand blustery conditions without budging. The barren soils of the desert, on the other hand, don't really have anything holding them firmly to the ground. Many years of weathering results in loose particles of sand and rocks on the surface, ready to blow around in whichever direction the wind carries them along.

## How Dust Storms Form

A powerful gust of wind across the desert floor can strip off the top layer of arid ground and send the particles flying into the air. Bigger, heavier grains of sand don't travel very far before gravity asserts itself and they fall back to the surface. Very fine particles of sand, called dust, don't succumb to gravity so quickly and can travel thousands of miles before eventually settling back to the ground.

Meteorologists use a few different terms to talk about major episodes of blowing dust. Dust storms and sandstorms can tower miles high into the sky and envelop a town for hours. Haboobs, an Arabic term that means "violent storm," are a particularly intense type of dust storm that are often associated with the outflow winds of a nearby thunderstorm. These three terms are often used interchangeably to refer to the same storm.

The most common triggers for dust storms are cold fronts and the winds of a thunderstorm outflow. Winds produced by fronts and thunderstorms are strong enough and last for long enough to build up a significant cloud of dust that can climb miles into the sky as it rolls downwind. The initial puff of dust helps scrape away the top layer and kick up more dust as the airborne particles blow along in the wind. One theory holds that the friction created by the blowing dust induces a static field that helps draw loose dust particles into the air, growing into a mighty cloud.

Haboobs are common in the American Southwest when severe thunderstorms bubble up during the rainy season. Powerful cold fronts rolling across the central plains can stir up dust on the flat lands of eastern Colorado and New Mexico and the panhandles of Oklahoma and Texas. Dust storms are a common threat in areas that have vast expanses of arid land, such as the deserts of Africa and Southwest Asia, western China and Mongolia, and even regions enduring a long-term drought that has desiccated and damaged the topsoil.

# HUGE HABOOB

One of the American Southwest's biggest haboobs in living memory swept across Phoenix, Arizona, on July 5, 2011. The dust storm measured more than 100 miles across and its dark embrace lasted for an hour or longer. The region hadn't seen meaningful rain in several months, providing the opportunity for a severe thunderstorm's powerful winds to scoop up plenty of dust and send it surging through the region.

# THE DUST BOWL

One of the most dramatic weather disasters in United States history unfolded at the intersection of extreme weather and extreme human mismanagement. An intense, long-term drought affected the southern plains states during the 1930s. The lack of precipitation dried out the ground, which helped intensify the drought by raising daytime temperatures, which reinforced the ridge of high pressure that prevented rainstorms from bringing any meaningful relief.

Droughts are bad enough on their own, but farmers across the region severely mismanaged their land and made a terrible situation downright disastrous. They plowed away the vegetation that held together the topsoil, allowing the earth to dry out when the drought hit. The combination of damage caused by

farmers and by drought destroyed the topsoil, turning it to infertile dust that blew into massive storms known as "black blizzards." Dust and loose dirt accumulated like snow in some communities, growing into dunes against barns and other objects exposed to the elements. The endless sight of desiccated land across the once-fertile plains gave rise to the "Dust Bowl" nickname that defines this era of quiet destruction.

The worst dust storm during the Dust Bowl era unfolded on April 14, 1935, a date ominously remembered as "Black Sunday." A stiff wind followed behind a cold front as it swept over the drought-stricken areas, whipping up a dust storm that darkened the sky for hours. The storm lofted so much dust and lasted for so long that residents feared the skies would never clear again.

## Dust's Effects on Weather

Blowing dust can influence weather conditions across areas as the particles flow downwind. Thick dust can darken the sky on a hot summer's day and serve to lower air temperatures for the duration of the event. A high concentration of dust in the upper atmosphere can enhance the greenhouse effect, temporarily increasing the amount of solar radiation trapped in the atmosphere.

The most prominent effect blowing dust can have on the weather comes courtesy of Saharan dust. Hot, dry winds blow an immense amount of dust off the Sahara desert and high over the Atlantic Ocean. The result is known as a Saharan Air Layer. Meteorologists can easily track a Saharan Air Layer on visible satellite imagery as the beige dust steadily makes its way across the Atlantic Ocean. Dry air within a Saharan Air Layer can suppress thunderstorm activity over the Atlantic Ocean, which can stifle hurricane activity during the height of the season.

One beautiful side effect of faraway dust storms are vivid sunsets. Folks in the Caribbean and eastern United States are often treated to brilliant sunrises and sunsets that fill the entire sky with color from one horizon to the other as a result of Saharan dust storms. Dust blowing in the wind at high altitudes is fine enough that it does a great job scattering sunlight as it passes through the atmosphere. When the Sun hangs low on the horizon at sunrise and sunset, light scatters through the dust and allows only the longest wavelengths to reach the surface. These long wavelengths—the reds, oranges, and yellows—provide a memorable scene courtesy of dust from thousands of miles away.

## Dust's Effects on Health

Blowing dust is fine enough to seep into cracks and crevices in your home, making it difficult to escape the grip of a powerful dust storm. Aside from a burst of allergies, folks who suffer from respiratory ailments such as asthma or chronic bronchitis may experience serious flare-ups during and after a dust storm that could require medical intervention. Dust storms are responsible for hundreds, if not thousands, of hospital visits every year.

Lowered visibility during a dust storm can wreak havoc on roadways. It takes only one car stopping to lead to a chain-reaction accident that could eventually involve dozens or even hundreds of vehicles colliding at high speeds. These incidents have led to an effort to send out emergency alerts to cell phones to warn people in the path of an impending dust storm.

# MARTIAN DUST

Earth doesn't hold a monopoly on blowing dirt. The atmosphere on Mars is just a tiny fraction as thick as Earth's atmosphere, but the winds that blow can stir up the fine Martian dust to dramatic effect. Mars's dust storms are so intense that they can completely encircle the planet, lasting for many weeks at a time before winds calm and the dust settles. NASA's Opportunity rover, which studied the planet for more than 14 years, met its final fate in 2018 during one of the most intense dust storms ever observed on the red planet. The voluminous dust likely covered the rover's solar panels, rendering it inoperable and unrecoverable.

# SUNRISE AND SUNSET

I f you want to appreciate the beauty of the skies
above, few moments beat sunrise and sunset.
No natural phenomenon is so romanticized
as a deep blue afternoon sky fading to the
chrysanthemum yellows and oranges, finishing
with a brilliant flash of red before the final hints
of a long day give way to night.

Without our atmosphere, dawn and dusk would be unremarkable. Think of the photos taken by the astronauts who traveled to the Moon. From their vantage point above Earth's atmosphere, the sky isn't even blue. Beyond the Earth, it is just the dark void of space. While not a shabby view, it highlights the artistic physics behind Earth's protective bubble.

The vivid reds and dulcet oranges that can illuminate the sky twice a day exist for the same reason that the sky is blue on a clear afternoon. Visible light is a form of radiation, much like the ultraviolet rays that can cause a sunburn. This radiation doesn't travel in a smooth line like a beam; instead, the radiation is wavy like ripples on a pond. The distance between each wave is known as the wavelength. The wavelengths that make up visible light are long enough that they don't injure us by attacking our cells, but they're short enough for our eyes to see them.

# RED SKY AT NIGHT, SAILOR'S DELIGHT

Accurate weather prediction is a relatively new practice. Humans had to develop inventive ways to predict the weather before we had access to technology. All the adages our grandparents used to talk about the weather were novel ideas back in the days before satellites and computer models. One of the best-known sayings from the days of yore is "red sky at night, sailor's delight; red sky at morning, sailors take warning." Where did we get such a lyrical forecast?

Believe it or not, the adage holds some water, and it speaks to the astute observations by folks like sailors whose lives and livelihoods relied on avoiding hazardous conditions and taking advantage of calm seas. Most weather systems in the Northern Hemisphere generally move from west to east. The rising sun in the eastern sky would cast its warmth on clouds and moisture moving in from the western sky, which could signal an approaching storm. The western sunset, on the other hand, would shine through the clearing sky to the west and highlight the clouds of a storm system moving away toward the east. The saying isn't always an accurate prediction, of course, but it's a nifty way folks could keep track of dangerous conditions in the days when storms were so often a surprise.

# GREEN FLASH

Witnessing a gorgeous sunset anywhere is a fulfilling experience, but there are few places that offer a better view of the waning moments of daylight than watching the Sun slip below the horizon over the ocean. Occasionally, if you're in the right place at the right time, you can witness a brief green flash at the top of the Sun.

What causes this elusive green flash? It's likely an optical illusion caused by a mirage over the ocean surface. A mirage occurs as a result of a temperature inversion, or when a layer of warm air covers a layer of colder air near the surface. Cold air is denser than warm air, so this inversion creates a sharp change in density in the atmosphere. This density change bends light back down toward the surface, allowing you to see objects and lights far beyond the horizon.

When the final seconds of direct sunlight shine through this temperature inversion, the very top of the Sun can appear green as a result of green light scattering through the inversion right above the surface. The phenomenon is much more pronounced on a clear, calm day, when a thick layer of cold air can develop over the water without interruption.

Visible light is made up of every color of the rainbow, collectively known as a spectrum. Each color has its own wavelength within the visible light spectrum. Blue light has a shorter wavelength than other colors like yellow and red, so it's easier for molecules in the air to scatter blue light and make the sky appear blue when we look up during the day.

As the Sun falls lower on the horizon at the end of the day or rises at the day's beginning, sunlight has to travel through a thicker portion of the atmosphere to reach the ground. It's like a molecular obstacle course that slowly filters out the blue light until all that's left are warmer colors with longer wavelengths. This is why the sky fades from blue to yellow, yellow to orange, and orange to red as the Sun sets, and the colors progress in the opposite direction during sunrise.

The colors at dawn and dusk are muted on clear, crisp days, and downright stunning on humid days or when the atmosphere is churning. It's not just the nitrogen and oxygen and other gases that scatter light. Even a modest collection of clouds at sunrise or sunset can fill the entire sky with a fiery display that no photograph can do justice. The effect is best when the Sun is low on the horizon, allowing the light to bathe the bottom of a high deck of textured clouds. The warm sunlight catches every bump and ripple in the base of the clouds, extending, for just a few brief moments, a sunrise or sunset that stretches as far as the eye can see.

It's not just clouds that can fill the sky with an unbeatable sight when the Sun hangs low on the horizon.. Haze, pollution, and wildfire smoke can also amplify the colors at dawn or dusk to almost surreal effect. While high clouds offer a pointed sight, a milky sky with little definition can lead to a diffuse palette that surrounds you with inescapable warmth. The effect is particularly otherworldly near an intense wildfire, where thick smoke is capable of filtering out almost all sunlight until the sky appears a dark shade of red.

The ash and gases released by volcanic eruptions can also lead to stunning, all-encompassing sunrises and sunsets around the world for weeks and even months at a time. The 1883 eruption of Krakatoa, a large volcano in Indonesia, released so much matter into the atmosphere that it reportedly painted skies around the world for months after the eruption.

> Don't forget:
> Beautiful sunsets
> need cloudy skies.
> —PAULO COELHO

# AUTUMN

The autumn months bear witness to all of nature's extremes in one short period of time. Fall begins with hurricanes and torrential downpours caused by summer's lingering heat, and the season ends with driving snows and surges of bitter cold plunging down from the polar latitudes. Crisp sweater weather and stunning foliage are the long-awaited reward of a tumultuous couple of months.

# AUTUMNAL EQUINOX

Don't you love the first brisk morning of fall? It's a joy to step outside and breathe in the sweet, metallic smell of fresh air after the season's premiere cold front ushers in cooler temperatures and lower humidity. While those relieving winds are a welcome change during the fall months, the weather rarely stays consistent. This muddled transition between summer and winter provides a little sample of every season, all wrapped into one three-month adventure.

The autumnal equinox marks the retreat of the Sun's direct rays out of one hemisphere and into the other. Despite the waning influence of the year's hottest sunshine, the atmosphere is in no hurry to cool down with the official onset of fall. Lingering humidity and summer-like weather patterns allow one warm day to bleed into the next even as the calendar flips along. It's not just the atmosphere that clings to its hard-earned warmth for as long as possible.

The long heat of a hot summer can make the oceans feel like bathwater. Sea surface temperatures in the Atlantic Ocean can climb to almost 90°F in parts of the Caribbean and Gulf of Mexico. The water retains heat after taking a long time to warm up. The Atlantic Ocean doesn't reach its peak heat until September, which coincides with the peak of hurricane season across the basin.

Sea surface temperatures in the Atlantic and Eastern Pacific stay warm enough to sustain tropical storm and hurricane activity through the end of November, with only a hostile atmosphere acting as a brake on further development. Many of the strongest hurricanes on record, such as Hurricane Ivan in September 2004 and Hurricane Wilma in October 2005, reached their peak during the fall months.

Even if you're lucky enough to avoid a hurricane, fall is still a rough season to live near the water. The fight between summer's stubborn grip and winter's impatient advance unfolds in dramatic fashion near oceans and big lakes. One infamous example unfolded on Lake Superior in November 1975. A freighter set sail from Wisconsin on November 9, 1975, to transport goods to ports around the Great Lakes. One day later, the ship got caught in a powerful storm and sank in the rough waters. The ship succumbed to the "gales of November," as Gordon Lightfoot put it in his hit song "The Wreck of the *Edmund Fitzgerald.*" The storm that claimed the now-legendary ship was a type of low-pressure system endemic to the fall months called a panhandle hook or, as meteorologists sometimes call it, a Texas hooker.

Cheeky nicknames aside, panhandle hooks are responsible for some of autumn's worst weather conditions in the central United States. These lows begin life on the eastern side of the Rocky Mountains near the Texas Panhandle and curve, or hook, toward the Great Lakes. The potent nature of these fast-moving systems can bring a community dangerous thunderstorms and muggy air during the day with a sharp drop in temperatures and driving snow just a few hours later.

Storms like panhandle hooks embody the unsettled conditions that blanket temperate regions in autumn. Fall weather patterns in the United States can produce just about every type of severe weather in just a few weeks. An active November could witness a hurricane in the Gulf of Mexico, a significant tornado outbreak in the Midwest, disruptive blizzards in the North, blustery winds and persistent storms raking the coasts, and warm conditions that spark horrendous wildfires in the West.

# WATCH YOUR BREATH

The hallmark of a cold day is the ability to see your breath when you first step outside. It's fun for kids (and adults!) to create clouds by simply exhaling into the frosty air. The water vapor in our hot breath condenses almost immediately when it comes in contact with our frigid surroundings, leading to a small cloud that rises up and away from our face. It's the same basic principle behind the formation of condensation trails, contrails, that develop in the moist exhaust of high-flying jet airplanes.

How long the cloud lingers depends on the humidity in the air. The cloud can last for just a split second when the air is dry, or it can last for a few seconds and drift away from our faces before the air we exhaled finally mixes with the environment. Folks who exercise on a subfreezing morning can even see frost develop on their eyebrows, eyelashes, or facial hair as the water droplets in their breath freeze on contact.

# CHANGING SEASONS
# CHANGE THE LEAVES

The skies are often a side attraction when it comes to fall weather in the United States. Hundreds of thousands of tourists flock to northern states in the heart of fall to take in the brief but spectacular display of colorful leaves that paint the region's hilly landscape.

Leaves on deciduous trees—like maples and oaks—change color as a direct result of the changing seasons. Tree leaves appear green during the warm season because they produce a pigment called chlorophyll, which is critical to the process of photosynthesis that converts solar radiation to the energy trees need to survive. When temperatures begin cooling down in the fall, which usually happens fast in the northern United States, trees cut off chlorophyll production to conserve energy so they can survive the winter.

Without any chlorophyll left to tint the leaves green, the leaves revert to their "original" colors, which are the vivid reds, oranges, and yellows we gawk at come the fall. Weather conditions during the summer and early fall can influence the color of the leaves. Warmer than average temperatures lasting through the early autumn can delay the change, resulting in duller colors that blend in with the scenery rather than pop out. This is why the colors are brighter in Michigan than in Mississippi, where it stays warmer longer. Drought conditions can damage the trees and force them to kill their own leaves in order to save the rest of the tree, causing the leaves to simply turn brown and fall off before they have an opportunity to shine.

Nor'easters begin forming along the Eastern Seaboard of the United States in October and November. While Halloween snowfalls aren't unheard of in the northeastern United States, nor'easters that spin up early in the autumn are hardly memorable because they're mostly rainmakers. Autumn nor'easters can wreak havoc on travel plans when they hit the week of Thanksgiving, buffeting major airport hubs with gusty winds, heavy rain, and low visibility, all making it difficult for airplanes to safely take off and land.

Lake effect snow also begins in earnest during the fall as cold air swoops down from Canada and rushes over the relatively warm lakes. Buffalo, New York, experiences their worst bouts of lake-induced snow occur during the fall.

One of the city's most impactful snowstorms began on November 17, 2014. A single band of lake effect snow formed just south of town and proceeded to drop nearly between four and six inches of snow per hour, resulting in 48 to 60 inches of snow falling in just a single day. The storm's timing, coming in the middle of fall, made the excessive snowfall even more disruptive because trees hadn't lost their leaves yet. The combined weight of heavy snow and leaves snapped trees and led to power outages.

# THE SOOTHING SMELL OF COLD AIR

Folks who favor chilly days love to take deep, refreshing breaths of cold air. But why does cold air seem to smell so good? The molecules that make up smells can float more freely when it's warm and humid outside. Our noses are also more sensitive to smells when we're surrounded by hot, moist air. That's why soap and shampoo have such a powerful smell during a warm shower, and it's why lawn trimmings and even dumpsters are so overwhelming when you step outside on a summer's day.

Your nose is less sensitive to smells when it's colder and drier outside. The smell and tingly feeling in your nose hits everyone differently—sometimes it's metallic, sometimes it evokes the sensation of chewing a mint or a menthol cough drop. You're actually detecting the *lack* of smell. The air smells fresh and clean because it's not packed with the scents and odors you typically detect on warmer days.

# EL NIÑO and LA NIÑA

Centuries ago, fishermen in Ecuador and Peru noticed that the water in the eastern Pacific Ocean was a little warmer than normal and that they weren't catching as many fish as they usually did for the season. Fish mysteriously vanishing is a big deal for towns and villages that rely on marine life for their economic survival. This phenomenon tended to appear around Christmas, so they named the event El Niño de Navidad after the Christ Child. Over the years, the name was shortened to El Niño, and scientists now recognize unusual temperature changes in this part of the Pacific as one of the most consequential events for seasonal weather around the world.

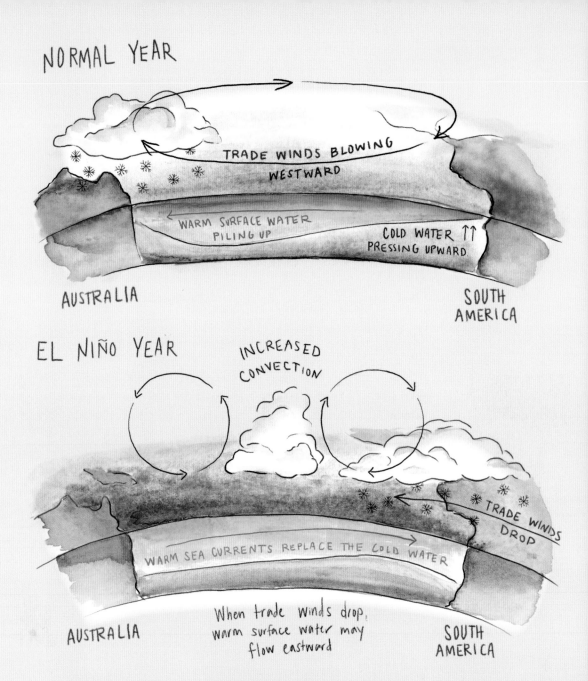

Winds blowing west across the Pacific Ocean sometimes weaken and allow warm water near Australia to flow toward South America, an abnormal event called an El Niño. When these easterly winds strengthen, cold water from deep within the ocean can rise and lead to unusually chilly water temperatures called a La Niña.

El Niño is one of those weather terms that elicits an immediate reaction even among folks who don't obsessively follow the weather's every twist and turn. The event is widely known for the wild effects it can have on faraway weather conditions, the name coming into widespread attention in the late 1990s after California witnessed tremendous rainfall and flooding during one of the strongest El Niño winters on record.

## Winds Break Down

An El Niño is a long period of abnormally warm water in the eastern Pacific Ocean near the equator. La Niña, which means "little girl" in Spanish, describes the opposite phenomenon, occurring when there's a long period of abnormally cool waters over the same stretch of the Pacific Ocean. Neutral conditions are present when water temperatures are near normal.

The official name for this phenomenon is the El Niño–Southern Oscillation, or ENSO for short. All these vast and dramatic changes in water temperature in the Pacific Ocean are rooted in the atmosphere above it.

A large-scale vertical air current called the Walker cell stretches more than 10,000 miles across the equatorial Pacific Ocean. The Walker cell forms from a sustained pattern of differing air pressures between the eastern Pacific and the western Pacific. Within the Walker cell, winds blow from a broad area of high air pressure near South America toward a broad area of low air pressure near Australia and Oceania, resulting in the prevailing winds that blow across the ocean.

These persistent easterly trade winds cause water on the surface of the Pacific Ocean to migrate from east to west. Since the wind is pushing surface water away from South America, cold water from deep in the ocean has to flow upward to take its place. This is called upwelling, and it keeps the eastern Pacific chilly and well-nourished—perfect for the Ecuadorian and Peruvian fishermen who earn a living off these bountiful waters.

Occasionally, the areas of high pressure and low pressure weaken or even reverse positions. A fluctuation in air pressure between the eastern Pacific and western Pacific is called the Southern Oscillation. Winds calm down when the air pressure gradient across the ocean weakens, which means that the flow of water from the eastern Pacific to the western Pacific also slows down or ceases. This allows all that built-up water near Australia, which has gotten quite toasty in the summer sun, to start flowing back toward South America. If this warm water sticks around for a while, it can turn into an El Niño event.

La Niña is roughly the opposite of an El Niño. Waters of the equatorial Pacific Ocean grow colder than normal when the Walker cell strengthens instead of weakens. Stronger high pressure over the eastern Pacific and deeper low pressure over the western Pacific enhances the easterly trade winds. Stronger winds push more water away from South America toward Australia, enhancing upwelling that causes sea surface temperatures to drop far below normal.

## Watching the Water Temperatures

Meteorologists monitor sea surface temperatures across the equatorial Pacific Ocean on a daily basis for signs that waters are growing warmer or colder than normal. Experts hone in on a 2.5 million square mile expanse of water that covers the span between 170°W and 120°W longitudinally, and from 5°N to 5°S latitudinally, an area that's simply referred to as "Niño 3.4."

Water temperatures in that Niño 3.4 region must register 0.5°C (32.9°F) above normal for about seven consecutive months in order to reach the criteria for official El Niño conditions. The same waters have to come in 0.5°C (32.9°F) below normal for the same amount of time for La Niña conditions to be present.

Just a one-half of one degree change in water temperatures in either direction doesn't seem like much, but the global climate is such a delicate system that even a minor change can throw everything off balance. Think of all the changes we've seen around the world as a result of climate change warming average temperatures by just a degree or two.

The change is just enough of a temperature anomaly over a long enough period of a time to have wide-ranging effects around the world. The potential for weather extremes is even greater when temperatures grow much warmer or colder than normal. The El Niño of 1982–1983 saw sea surface temperature anomalies as much as 7.2°F above normal off the coast of Peru. The memorable El Niño event of 1997–1998, the one that influenced that winter's floods in California, saw a widespread area of temperatures more than 5.4°F above normal.

## El Niño Winters

Aside from leading to economic difficulties for residents of western South America, El Niño has a long reach that touches just about every country that borders the Pacific Ocean. Sea surface temperatures can influence weather patterns above the water.

We're most familiar with the event's influence on winter weather in the United States. El Niño winters in the US are generally warm across most of the country, with wetter-than-average conditions across the southern states and drier-than-normal conditions in the northern states.

The deviation from the wintertime normal in the US is a direct result of how El Niño alters the jet streams that move over the country during this part of the year. When the ocean grows warmer as an El Niño settles in, the air above the ocean grows warmer in response. Warming air over the eastern Pacific Ocean causes the atmosphere to expand, which leads to a ridge of high pressure over the region. This ridge of high pressure strengthens the subtropical jet stream that moves over the southern United States.

A stronger subtropical jet leads to low-pressure systems at the surface that move over California and continue tracking east until they reach the Atlantic coast. These are the lows responsible for drenching California with more rainfall than the state's waterways can handle. Meanwhile, up north, the same ridging that strengthens the

# TELECONNECTIONS

El Niño–Southern Oscillation (ENSO) is a fantastic example of a teleconnection, or how an unusual weather pattern or other anomaly (like warmer waters) in one part of the world can affect weather conditions many thousands of miles away.

Teleconnections offer important clues to meteorologists looking to understand the driving forces behind weather conditions in a certain region. Monitoring teleconnections gives you a general idea of what kind of conditions to expect in the following weeks and months. In addition to ENSO, another well-known teleconnection is the Arctic Oscillation (AO), which is associated with the polar vortex that can produce deep cold snaps far south of the Arctic Circle.

subtropical jet stream forces the polar jet stream to bend north into Alaska and Canada. When the polar jet juts north, warmer conditions blanket the northern states, and the jet's position prevents winter storms from walloping the region as frequently as they would during a normal winter.

The same ridging that brings topsy-turvy conditions to the United States keeps much of the western Pacific warm and dry during the winter season. India, the countries of Southeast Asia, and Japan typically experience a warm winter during El Niño.

On the other side of the equator, El Niño winters are relatively boring in Australia and Oceania. Since the region's warm waters migrated eastward to set the whole process in motion, the southwestern Pacific is left cooler and drier than normal. This can lead to a chilly winter in places like Sydney, Australia, and an intense El Niño can even set off a prolonged drought that increases the chances for wildfires and taxes the region's freshwater supply.

## La Niña Winters

While a La Niña is close to the opposite of an El Niño, the effects of the cooler water on global weather patterns aren't simply the reverse of what happens during an El Niño. The atmospheric cooldown that corresponds to the cooler waters weakens the temperature and pressure gradients that would otherwise allow the jet stream to strengthen and make conditions active and interesting.

Across the United States, a winter dominated by the effects of La Niña sees warmer than normal conditions across the southern states, similar to an El Niño, but without much of a difference in precipitation patterns either way. Western Canada and the northwestern United States typically see a cool winter during La Niña, a result of the polar jet stream sagging farther south than normal.

During winter in the Southern Hemisphere, cool conditions dominate western South America due to strong upwelling of the eastern Pacific Ocean, while Australia and Oceania bask in warmer than normal wintertime temperatures thanks to the surplus of warm water shoved their way by the stronger easterly winds.

## El Niño Summers

Summertime isn't as widely influenced by El Niño as the winter, but it does create one major influence that is felt all summer long. Atlantic hurricane seasons are greatly subdued during an El Niño summer or fall. It seems strange that hurricane activity in one ocean basin is influenced by the water temperature in another basin, but it's all connected.

Warmer air temperatures over the eastern Pacific Ocean increases the amount of wind shear that flows east toward the Caribbean and the western Atlantic Ocean. Wind shear is toxic to a budding tropical storm or hurricane because it shreds the tops off the thunderstorms before they have a chance to develop and take root. Persistent wind shear influenced by the warmer-than-normal Pacific can lead to far fewer Atlantic hurricanes than we'd see in a normal season.

## La Niña Summers

A La Niña summer is something to worry about if you live near the coast of the Atlantic Ocean. While variations in winter weather aren't direct opposites from an El Niño to a La Niña, the two events do have mirrored effects on the Atlantic hurricane season that can lead to tragic consequences.

Cooler temperatures over the eastern Pacific Ocean allow winds to calm down, which means that there's little to no destructive wind shear blowing toward the Atlantic Ocean to put a lid on any hurricanes that might form. As a result, La Niña conditions during the summer and fall months tend to allow the Atlantic hurricane season to flourish.

A moderate La Niña event helped make 2020 the most active hurricane season on record in the Atlantic Ocean. That year produced an astounding 30 named tropical storms or hurricanes, more than doubling the basin's annual average number of named storms.

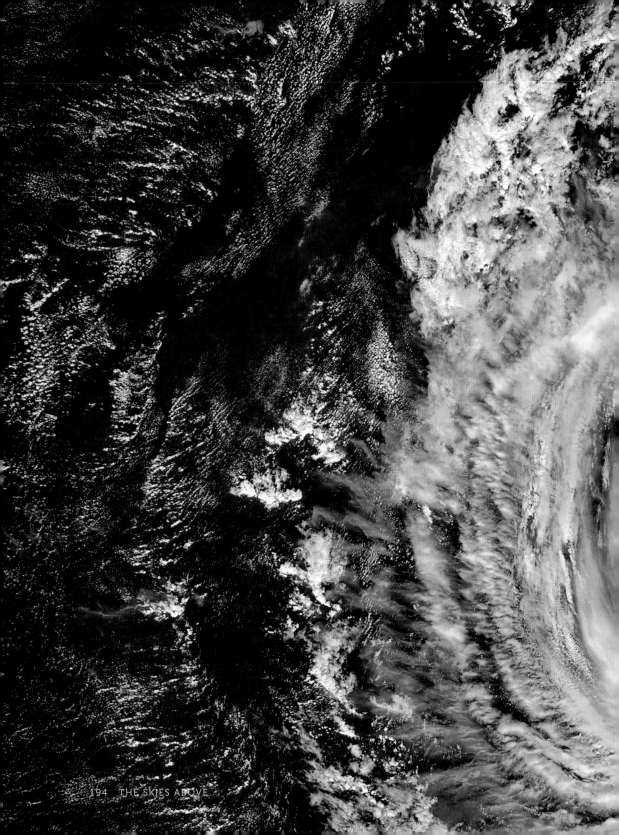

# HURRICANES

**C**oastal communities offer some of the best scenery and relaxation the natural world can offer. Soaking up the warm sunshine on a pristine beach while a cool breeze blows over crashing waves is the ultimate goal for folks who dream about taking some much-deserved time off. But life isn't always grand when you live near the water. Every warm season, just as folks flock to the coast to enjoy a quick vacation, danger can spin up and roar ashore with frightening intensity. Tropical cyclones are among the strongest storms on Earth, and their relentless winds and horrifying flooding can wash communities away in a flash.

## Different Oceans, Different Terms

The term "tropical cyclone" is the catch-all name for a low-pressure system that displays tropical characteristics. Tropical cyclones draw their energy from thunderstorms near their center of circulation, and they have warm and muggy air throughout the entire storm.

A tropical depression, tropical storm, and hurricane all represent different stages of tropical cyclone development.

- **TROPICAL DEPRESSION:** A tropical depression is a budding tropical cyclone that's disorganized, weak, and ragged in appearance.

- **TROPICAL STORM:** A tropical storm forms when a tropical depression's minimum air pressure drops and its winds grow stronger. A tropical depression becomes a tropical storm when its maximum sustained winds reach about 40 miles per hour.

- **HURRICANE:** A tropical storm becomes a hurricane when it develops a strong core of thunderstorms around the center of circulation, intense rain bands that spiral away from the storm, and often a clear eye that forms at the center of the storm.

Mature tropical cyclones are called different names around the world. We call them hurricanes when they form in the Atlantic Ocean or in the central or eastern Pacific. These systems are known as typhoons in the western Pacific Ocean, while they're simply called cyclones in the southwestern Pacific and the Indian Ocean.

Tropical cyclones are common around the world during each region's warm season. Most tropical cyclones that form in the Atlantic Ocean and much of the Pacific Ocean occur between the late spring and the late autumn months, peaking with the ocean's peak heating at the end of the summer. The climates of Southeast Asia and the Indian Ocean are warm enough that tropical cyclone seasons there have no official bounds, with countries like the Philippines at risk for tropical cyclones year-round.

Colder waters, though, almost never see tropical activity in their own right. A map of all tropical cyclones tracked over the past two centuries reveals a few conspicuous quiet spots around the world. The waters are too cold and the climate is too hostile to support tropical cyclone development in the Arctic and Southern Oceans. Waters are also too chilly off the western coast of South America, where no tropical cyclones have ever been observed. Meteorologists have tracked only a handful of storms in the South Atlantic Ocean between South America and Africa. Several tropical storms have formed there over the years, but the basin has recorded only one hurricane—Hurricane Catarina, which hit Brazil in March 2004—since the beginning of the satellite era in the 1960s.

## Names Are a Big Deal

Tropical cyclones are big enough and common enough that it's cumbersome to track them all, especially if two storms form close together and threaten the same general area. Beginning in the mid-1900s, meteorologists began naming tropical cyclones to keep track of the systems and to help communicate their threats to the public.

These naming schemes are maintained by the World Meteorological Organization (WMO), the agency of the United Nations that maintains worldwide weather standards. Each ocean basin has its own lists of names and its own naming schemes. Responsibility for issuing names and keeping track of storms falls to each region's designated weather bureau, which is the US National Hurricane Center for the Atlantic, the eastern Pacific, and the central Pacific around Hawaii.

Each ocean basin's names reflect the languages and cultures of each region. The Atlantic Ocean's names all derive from English-, Spanish-, and French-speaking cultures, while the western Pacific's names are contributed by the countries of East Asia.

The National Hurricane Center draws names for storms that form in the Atlantic basin from one of six different lists. One list of names is used every six years, so 2022's storm names were last used in 2016, and the same list won't come into use again until 2028. Each list contains 21 names in alphabetical order, and the names alternate between masculine and feminine. The Atlantic basin's lists exclude the letters Q, U, X, Y, and Z for lack of common names that could be used as suitable replacements if a certain name is retired.

A list of 21 names is more than enough to cover a typical hurricane season in the Atlantic Ocean, which experiences about 14 named storms in an average year. However, on two occasions, the Atlantic produced more than 21 tropical storms. When forecasters exhaust a season's official list of names, they used to use the letters of the Greek alphabet

# RETIRING NAMES

Hurricane names live on in the minds of survivors long after a destructive storm is gone. Just mentioning the name "Katrina" in Louisiana brings up horrible memories for folks who lived through that awful storm and its aftermath in August 2005. Meteorologists are sensitive to the cultural and psychological impact of highly impactful storms, so they "retire" the names of storms that cause large numbers of deaths and vast destruction so that particular name won't be used again.

The WMO has retired and replaced dozens of storm names from the official lists used in the Atlantic and eastern Pacific over the past couple of decades. When a name is retired, officials pick another name beginning with the same letter to serve as a replacement.

Hurricane Ivan devastated parts of the US Gulf Coast in 2004, forcing officials to replace "Ivan" with "Igor" for the 2010 Atlantic hurricane season. They promptly had to retire the name "Igor" after the hurricane that earned the name's inaugural use damaged Newfoundland in eastern Canada. Officials then replaced "Igor" with "Ian" for the 2016 hurricane season.

as a fallback to name any additional storms. The 2005 hurricane season required the use of six Greek letters to name the season's final storms, while the 2020 hurricane season required the use of nine Greek letters.

The WMO decided to stop using the Greek alphabet as a backup naming scheme beginning with the 2021 hurricane season because the system proved confusing and distracting when multiple serious storms approached land. The agency developed a supplemental list of common names (such as Caridad and Gerardo) to use for any future hyperactive hurricane seasons.

## Condensation Warms Things Up

Every tropical cyclone begins with a seed. The seed can be a decaying cold front, a trough of low pressure, or a complex of thunderstorms that starts over land and moves over the ocean. The latter is the most common way a classic hurricane in the Atlantic Ocean comes to life. Late in the summer, during the height of sub-Saharan Africa's monsoon season, groups of thunderstorms persistently move off the coast near the Cabo Verde Islands and seed the development of a future storm.

The key to a tropical cyclone's development is sea surface temperatures. Water temperatures generally have to be 80°F or warmer in order to support tropical cyclone development. Sea surface temperatures up near 90°F, which are common late in the summer in parts of the Caribbean and the Gulf of Mexico, can support some of the strongest hurricanes on Earth.

Thunderstorms normally develop when warm air near the surface rapidly rises through cold air above. That doesn't quite happen in the tropics, where there's not a sharp temperature gradient between the lower levels and the upper levels. That's where warm ocean temperatures play a role.

Water vapor has the ability to both add and subtract heat from the atmosphere. Evaporation is a cooling process. That's why we cool off when we sweat. Condensation is a warming process. Water vapor releases heat into the atmosphere when it condenses into a cloud. This process is called latent heat release, and the warmth released by water vapor condensing into clouds is how thunderstorms gather the instability they need to strengthen and organize over tropical waters.

A group of thunderstorms can feed off of warm ocean waters by taking advantage of latent heat release to keep the storms healthy and animated. The gusty winds of a vigorous thunderstorm help evaporate some of the hot water off the surface of the sea, sending that water vapor skyward to condense and continue fueling the thunderstorms above.

## How a Tropical Cyclone Forms

If these thunderstorms are strong and persistent enough, the updrafts will remove a massive amount of air from the surface and send it high into the atmosphere. This leaves an area of low pressure at the surface. If thunderstorms persist around this newly formed center of low pressure, and the storms contain sustained winds of about 30 miles per hour, meteorologists will call the system a tropical depression.

A tropical depression needs favorable conditions to continue strengthening. Warm water is only a part of the equation. Tropical cyclones struggle if there's too much wind shear throughout the atmosphere. Speedy winds a few thousand feet above the surface knock over a thunderstorm's updraft as it tries to get organized, shredding a storm apart as if it were made from cotton candy. Even a strong tropical cyclone can fall apart in just a couple of hours in the face of strong wind shear.

A tropical cyclone also needs moist air to feed the storms. Dry air can starve a tropical cyclone's core thunderstorms and force the system to weaken. This is a big problem for storms that form late in the season, when cold fronts start moving over the open ocean, or when Saharan dust storms travel west across the Atlantic.

If all of those conditions are present—warm water, moist air, and calm winds—a system can continue to develop as long as its internal structure remains organized. As thunderstorms strengthen around the center of low pressure at the surface, the minimum central pressure will fall and the winds will pick up in intensity in response.

## The Storm Strengthens

A tropical depression strengthens into a named tropical storm when its maximum sustained winds grow to 40 miles per hour. The system will begin to take on a classic swirly appearance on satellite and radar imagery as bands of showers and thunderstorms spiral around the system like spokes on a wheel.

Once a tropical storm's winds reach about 75 miles per hour, the system is officially a hurricane. The air pressure at the center of a hurricane drops low enough that nature is desperate to correct the enormous imbalance. Air rushes in from the sides and top of a hurricane to try to fill the low and balance out the atmospheric pressure again.

Air sinking through the center of a hurricane warms up and dries out as it's pulled to the surface, resulting in the clear eye that makes these storms so memorable and ominous. The eye of a hurricane isn't always perfectly symmetrical. Depending on the storm's structure and the conditions around it, a hurricane's eye can fill with clouds or appear completely clear, and the eye can measure dozens of miles wide or so small that you can barely spot it in satellite imagery.

While some air near the top of the hurricane sinks back through the eye of the storm, much of the air vents out and away from the storm like an exhaust fan. This outflow of air keeps the hurricane healthy and thriving. Thin bands of rippling cirrus clouds often form in the moist winds blowing away from the top of a hurricane, leading to the classic "buzz saw" appearance that makes these storms look so ominous from above.

## Powerful Winds

Hurricanes can grow incredibly powerful. The strongest hurricane ever recorded (in terms of sustained winds over a period of one minute) was Hurricane Patricia in the

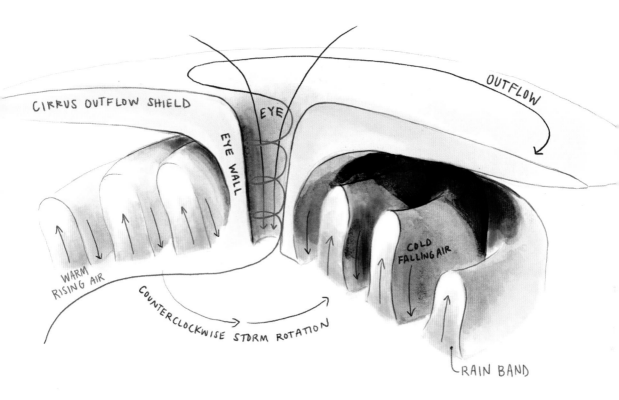

eastern Pacific. Hurricane hunters measured 215-mile-per-hour winds near the eye of the storm as it approached the west coast of Mexico on October 23, 2015. The system weakened a little bit before coming ashore north of Manzanillo, Mexico, with 165-mile-per-hour winds, still making the storm one of the most intense hurricanes ever observed at landfall in the country.

Meteorologists often use the Saffir-Simpson Hurricane Wind Scale to convey the strength of a hurricane. The scale measures winds ranging from category one to category five, with higher categories indicating higher intensity. Only a handful of storms around the world ever reach or exceed 160 miles per hour to become scale-topping category five hurricanes.

Use of the Saffir-Simpson scale is highly controversial among meteorologists due to the issues the scale causes when a storm approaches the coast. A category one hurricane can cause as much death and destruction as a category three hurricane, depending on its size and where the storm hits.

# HURRICANE SANDY:
# THE UNUSUAL SUPERSTORM

Hurricane Sandy carved out a unique spot in weather history when it rolled ashore near Atlantic City, New Jersey in late October 2012. Even though the storm was "only" a category one with maximum winds of about 85 miles per hour when it made landfall, Sandy's incredible size left a lasting footprint of destruction on the northeastern United States.

Low-pressure systems exist on a spectrum and aren't always easy to classify as purely tropical or purely extratropical storms. Sandy is the epitome of a storm that's hard to fit in one box. While this hurricane formed from tropical roots over the Caribbean Sea, the storm felt the effects of a powerful jet stream as it approached the United States. As Hurricane Sandy raced toward the cooler waters and hostile winds in the northwestern Atlantic and started to lose its tropical characteristics, the jet stream supplemented its lost power and helped the storm grow.

By the time Sandy reached landfall in New Jersey, the system was no longer considered a hurricane due to its transformation. However, its effects were extraordinary, generating tropical storm force winds across an area that measured nearly 800 miles wide, affecting 24 states from Florida to Maine. This broad expanse of wind pushed a tremendous storm surge into the New Jersey and New York coastlines, producing widespread flooding on top of extensive wind damage. The storm was so large that when it met with the cold air in the Mid-Atlantic states, it produced blizzard conditions in the Appalachian Mountains.

# THE SAFFIR-SIMPSON SCALE

The Saffir-Simpson Scale is widely used to give folks in harm's way a quick idea of the magnitude of a hurricane's strongest winds. The scale has five categories.

- **CATEGORY ONE (74–95 MPH):** Many trees will fall under the stress of the winds. Homes could sustain minor damage.

- **CATEGORY TWO (96–110 MPH):** Homes and businesses will begin to suffer major roof damage. Widespread power outages will result from trees falling on power lines.

- **CATEGORY THREE (111–129 MPH):** Buildings will begin to suffer major structural damage, including total roof loss and wall damage.

- **CATEGORY FOUR (130–156 MPH):** Many homes and businesses may be destroyed.

- **CATEGORY FIVE (157+ MPH):** Damage is so thorough and widespread that forecasters usually warn that areas affected by the strongest winds will be "uninhabitable for weeks or months" after the storm.

Lots of coastal residents think, "It's *just* a category two" when they're deciding whether or not to evacuate. Widespread misconceptions about a storm's true hazards caused by the Saffir-Simpson scale are a huge problem for meteorologists and local officials trying to prepare for a storm. The hesitation to evacuate based on a hurricane's category has resulted in unnecessary suffering, injuries, and deaths for folks who chose to stay behind and for the rescuers who have to wade into danger to save them.

## Flooding and Tornadoes

Intense winds are only one part of a tropical cyclone's hazards. Wind is dangerous, but water is responsible for most of the deaths during a landfalling storm. A storm surge occurs when a storm's strong winds push seawater into the coast. Strong hurricanes

# HURRICANE HUNTERS

"Hurricane Hunters" is a common name given to groups of scientists who fly specially equipped planes into the heart of a hurricane to directly sample the core of the storm and help forecasters send out accurate warnings and forecasts to folks in harm's way. Aircraft reconnaissance missions are usually flown by experts working for NOAA and the US Air Force.

Reconnaissance aircraft are equipped with the ability for crew members to release radiosondes—called "dropsondes" when they're dropped from the belly of the plane—into the storm below, taking wind, pressure, temperature, and moisture readings to gauge the true strength of the storm. One of the specialized instruments aboard the plane is called the Stepped Frequency Microwave Radiometer (SFMR). This sensor monitors the ocean below the aircraft and observes the sea foam blowing around on the surface of the water, using this data to accurately determine the wind speeds at the surface.

can generate a storm surge more than 10 feet deep, completely inundating buildings near the coast and sometimes pushing far inland from the beach. Much of the destruction wrought by Hurricane Katrina in 2005 and Hurricane Sandy in 2012 resulted from their storm surges.

Freshwater flooding from heavy rain is a major threat with any landfalling storm. A slow-moving tropical cyclone, even one as weak as a tropical depression, can produce more than a foot of rain in extreme cases. This much heavy rain falling this quickly can lead to widespread flooding that can submerge areas that typically don't experience any issues during normal thunderstorms.

The geography and weather patterns of the southern United States make the region exceptionally vulnerable to flash flooding from tropical cyclones. Tropical Storm Allison (2001), Hurricane Matthew (2016), Hurricane Harvey (2017), and Tropical Storm Imelda (2019) are all examples of storms that led to devastating flooding in the southern United States.

One dangerous and often overlooked hazard from a tropical cyclone is tornadoes. Strong wind shear in the lower levels of a landfalling storm is ripe for thunderstorms to begin rotating and producing tornadoes. Tropical tornadoes are a unique danger because they can form so quickly that forecasters have only a minute or two to issue tornado warnings, and some tornadoes can completely evade radar detection by forming so low in the atmosphere that the rotation occurs beneath the height of the radar beam.

# FOG

We've all had mornings where we peek out the window and see hardly anything at all, but after a few blinks realize that it's just really foggy. Fog can be beautiful, such as when you crest a hill and gaze down at a valley full of smooth clouds hugging the ground, gently glistening in the morning sun. But low visibility is a serious threat to your safety when you're traveling at high speeds and can't see more than a few feet in front of you. Fog, while common, is a perfect example of how tiny changes in our surroundings can make a huge difference in the weather around us.

## How Fog Forms

Fog is essentially a stratus cloud that forms at ground level. The most important factor in fog formation is the humidity. The dew point is the best metric to judge the amount of moisture in the air, but relative humidity—that old measurement that's ubiquitous in weather reports—is pretty useful when it comes to predicting fog. The air has to come close to complete saturation, or approach 100 percent relative humidity, in order for water vapor near the surface to condense into fog.

# Fog is rain that whispers.

—OLIVIA DRESHER

## Radiation Fog

Temperatures cool off fast when the Sun sets. It's not uncommon for temperatures to fall more than 10°F within an hour of sundown. The Earth's surface has a low heat capacity, which allows dirt and pavement to warm up quickly during the day and radiate that heat back into the atmosphere as soon as the Sun goes down. Clouds act like an insulator that keeps heat close to the ground, but if the skies are clear, the heat can escape and cool off a young night in a hurry. This is called radiational cooling, and it's responsible for some of the most photogenic fog you'll encounter.

When radiational cooling occurs on a night when there's plenty of moisture in the air, the air can reach 100 percent relative humidity. Once the air hits its saturation point, a thin layer of fog develops in the coolest air right above the surface, slowly growing thicker as the air cools down with height.

If you've ever seen an open field shrouded by a layer of fog that's only a few feet thick, almost like it's a special effect in a scary movie, then you've witnessed radiation fog. This kind of fog tends to be shallow. It's not uncommon to be able to look up and see the Moon and some stars through the top of the fog even if it's tough to see more than a few dozen feet around you.

Radiation fog commonly "burns off" within a few hours of sunrise. Bright early-morning sunshine quickly raises air temperatures from their predawn low, which lowers the relative humidity and causes fog to dissipate. It's common for radiation fog to dissipate from the top down, allowing folks on hills, bridges, and even the upper floors of buildings to find themselves above the layer of fog as the sunshine warms the air.

## Advection Fog

Sunrises and sunsets aren't the only way the atmosphere warms up or cools down. Air is in constant motion. Meteorologists use the term advection to refer to air or moisture that moves from one place to another. Chilly air swooping in behind a cold front is a form of cold air advection, while a sea breeze that makes the afternoon muggy is an example of moisture advection.

Fog is a common byproduct of cold and warm fronts moving through a region. Advection fog forms when cold air moves into a region of warm, muggy air, or when humid air shoves up against a cooler air mass. Warm fronts swirling around low-pressure systems can crash into a region of cooler air nearby. The warmer air suddenly cools as it mixes with the colder air, quickly bringing the air to its saturation point. If it's a bit chilly out and the forecast calls for severe thunderstorms, for instance, you'll know moisture is moving in and you're at risk for strong storms when it suddenly gets foggy.

Advection fog is a big problem for folks who live in coastal communities where the waters get quite chilly for a portion of the year. Fog is a frequent visitor in areas like California's San Francisco Bay, where the waters are always cool, and in the northeastern United States during the early spring months when warm air flows over chilly waters. The cool water temperatures lower the temperature of the air directly above, bringing the air to its dew point and causing a thick layer of fog to develop above the ocean's surface. This fog can roll inland and bring poor visibility to coastal areas.

## Valley Fog

The valleys between large mountains are the perfect shape for dense fog to settle in and stick around for hours or even days when weather patterns are stagnant. Cold air draining down the slopes of mountains can get trapped in valleys if there isn't any wind to mix the air up. This leads to a strong inversion, or cold air capped by warm air above, that traps the chilly air at the bottom of a valley.

Fog isn't the only problem in valleys. Cities like Salt Lake City, Utah, and Beijing, China, are situated in valleys that often suffer from inversions that allow for many days of pollution-laden fog to settle during the colder months. This soupy, polluted air mass can trigger major health issues for folks who suffer from asthma and other chronic respiratory issues.

## Snow, Hail, and Freezing Fog

Frozen precipitation can lead to localized areas of dense fog. If warm temperatures follow a recent snowfall, the air right above the snow can quickly cool and condense, forming a layer of fog that's only a few feet thick. The same phenomenon can happen after a severe thunderstorm on a hot day. If a thunderstorm generates so much hail that it accumulates like snow, air temperatures can plummet and make it difficult for folks to navigate the ice-covered roadways.

Fog is especially dangerous when it forms as temperatures fall below freezing. Supercooled water droplets suspended in the air can freeze on contact with any exposed surfaces outdoors, leaving behind a thin film of ice on roads, sidewalks, and vehicles. The ice that results from freezing fog can turn out to be as impactful as a coating of ice from freezing rain.

## Fog's Hidden Dangers

Poor visibility caused by fog is responsible for some of the worst transportation disasters in modern history. Every year, the United States experiences dozens of serious pileup accidents on busy highways when thick fog causes an accident that oncoming drivers can't see until it's too late.

One of the United States' deadliest train crashes took place near Mobile, Alabama, on September 22, 1993. A towboat got lost in the fog in the middle of the night while traveling up the Mobile River, accidentally turning into Big Bayou Canot and striking a railroad bridge. The collision buckled the tracks, causing an approaching passenger train to derail in an accident that killed nearly 50 people.

Flying is extremely tricky when fog is present at takeoff or landing. The worst plane crash in aviation history claimed more than 500 lives in 1977 when two Boeing 747s collided on a fog-covered runway on Tenerife, part of the Canary Islands in the northeastern Atlantic. Modern aircraft are equipped with advanced technology to help pilots see what's ahead of them even if they can't see out the windows, but pilots are trained not to fly if the visibility drops below a certain level.

# THE BLUE OF THE BLUE RIDGE MOUNTAINS

The Blue Ridge Mountains are a popular destination for vacationers in the eastern United States looking for a quick mountain getaway. Part of the Appalachian Mountains, stretching from northern Georgia to southern Pennsylvania, the region got this nickname from the bluish haze that covers the region's flowing terrain on a hot summer's day.

While it can look like fog, the haze over the Blue Ridge earns its faint blue color as a result of a chemical compound called isoprene. The trees release it on the hot and humid days that smother the eastern United States during the summer. Sunlight shining through the haze hits the isoprene and scatters the blue light, lending the Blue Ridge the iconic tint.

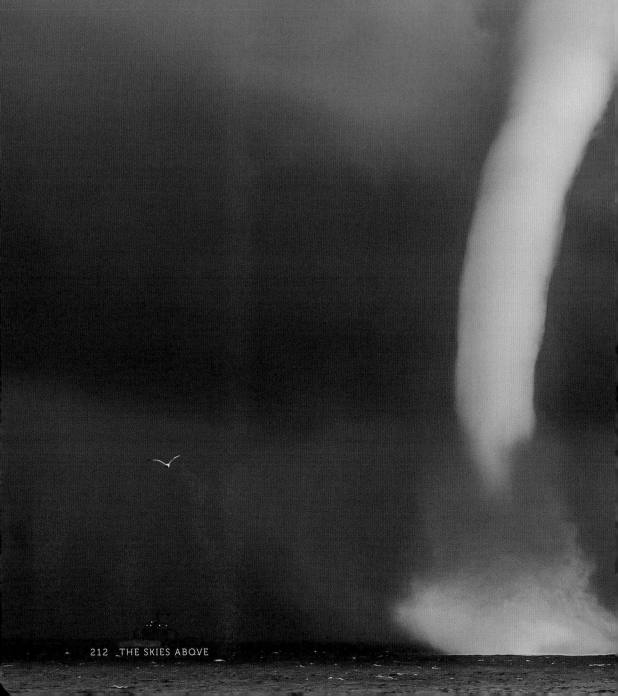

# WATERSPOUTS, DUST DEVILS, and FIRE WHIRLS

Tornadoes aren't the only way that our skies can begin twirling—there are a bevy of assorted air whirls that can turn a serene day on its head. Waterspouts dance off the coasts of the world with unsettling frequency. Dust devils tower high into the hot afternoon sky over deserts and parking lots alike. And some of the fiercest wildfires can maximize their heat and fury by forming a fire whirl that can compound the damage on an already scorched landscape.

# Waterspouts

Waterspouts are one of those forces of nature that are somewhat safe to admire from a smart distance. This rotating column of air forms over open water instead of land when temperatures are warm, there's plenty of moisture in the air, and winds are relatively light. There are two types of waterspouts: fair-weather waterspouts and tornadic waterspouts.

Even though they're exceptionally common, the exact mechanism that causes a waterspout to form is still up for debate. Fair-weather waterspouts typically form beneath growing cumulus clouds. The most common theory is that the rising air feeding the cumulus cloud's formation can begin to spin if there's a little bit of light wind shear from a sea breeze or the winds from nearby showers or thunderstorms. This rising air begins to spin faster as it rises toward the cloud and stretches out, turning into a waterspout. While these whirlwinds form anywhere that gets warm enough to see regular cumulus clouds, they're a frequent sight around the Gulf of Mexico, the Florida Straits, and in warm climates like the southern Mediterranean Sea.

Contrary to popular belief, a waterspout isn't a solid column of water stretching out of the ocean. We see a visible funnel stretching from the clouds to the water because of water vapor condensing in the lower air pressure within the waterspout itself. The relatively gentle motions inside a waterspout (compared to the violent winds of a tornado) make these formations appear smooth and almost frozen.

Usually, fair-weather waterspouts come and go without much fuss, but sometimes the winds can grow as strong as 60 miles per hour. These speeds can threaten to capsize smaller boats and are a danger to folks on the beach and close to the coast. Even though waterspouts are relatively weak, they can pick up debris like umbrellas, furniture, and trash cans and toss them through the air at dangerous speeds. Forecasters often issue a tornado warning if they know a waterspout is about to come ashore.

Tornadic waterspouts, as the name suggests, are simply tornadoes that form over the water. Common in places such as the Gulf of Mexico, where supercell thunderstorms develop close to the coast during a tornado outbreak, tornadic waterspouts form through the same process that produces a tornado over land, and they can be just as deadly.

# LANDSPOUTS AND GUSTNADOES

Some severe thunderstorms are capable of producing tornado-like whirlwinds known as landspouts and gustnadoes. Landspouts are tall, smooth tubes of rotating air that commonly form in dry, high-elevation areas like eastern Colorado and the Great Plains. Landspouts are closely related to waterspouts in how they form, since they aren't produced by a rotating updraft like a true tornado.

Gustnadoes are small whirlwinds created by the strong wind shear of a thunderstorm's downdraft racing away from the base of the storm. A gustnado is similar to the little whirlpool that forms when you run your hand through a swimming pool. While neither landspouts nor gustnadoes are "actual" tornadoes, they can still produce dangerous wind damage and should be treated as seriously as tornadoes when they approach.

## Dust Devils

Spend time near a dirt field, a big farm, or a seemingly endless parking lot, and you'll probably witness at least one dust devil in your life. This rapidly rotating column of air forms over hot surfaces and kicks up whatever dirt and particles are in its path.

Dust devils can develop on sunny, breezy days. Bright sunshine doesn't heat the surface at an even pace. Some spots end up a bit cooler than others, which affects where air can begin to rise through convection. If the air over a parking lot heats up enough through contact with the scorchingly hot surface, for instance, the air can quickly start to rise. The rising air can begin to rotate as a result of gusty winds. This rotating column of air begins to stretch upward as the air rises higher into the sky, rotating faster and causing a dust devil.

The appearance of a dust devil depends on what kind of ground the whirlwind traverses. If the rotation begins over a roadway, it'll probably appear a whiteish-tan color as it sweeps up dirt that's accumulated on the pavement. Dust devils that form in the Australian Outback are deep red because of the oxidized sands that blanket the center of the continent. These formations don't only pick up dust and dirt. "Hay devils" are common on farms when a dust devil moves over loose hay, while folks visiting ski resorts can occasionally witness a rare snow-filled dust devil that forms on the slopes.

Unlike waterspouts, dust devils are tiny but fierce. Most dust devils measure only a few feet across and are a little taller than the average human at their biggest. They're a novelty to look at, and the smaller ones are even a little fun to run through if you don't mind having to blow your nose for a few days after. But the winds of a strong dust devil can make you lose your balance, and they pick up gravel, dirt, and debris, which can cause abrasions.

Some dust devils, such as those that form in the desert, can extend as high as a mile in the sky and grow quite powerful. The strongest dust devils can rival the strength of a weak tornado, damaging small buildings like sheds and barns, and sending relatively heavy debris like lawn furniture and trampolines soaring into the sky. These dust devils are dangerous, and they've caused plenty of injuries (and even some fatalities!) in the past.

## Fire Whirls

Wildfires are intense. The worst forest fire can burn at more than 1,000°F, which is hot enough to melt the windows and hubcaps right off of a car. The massive amount of

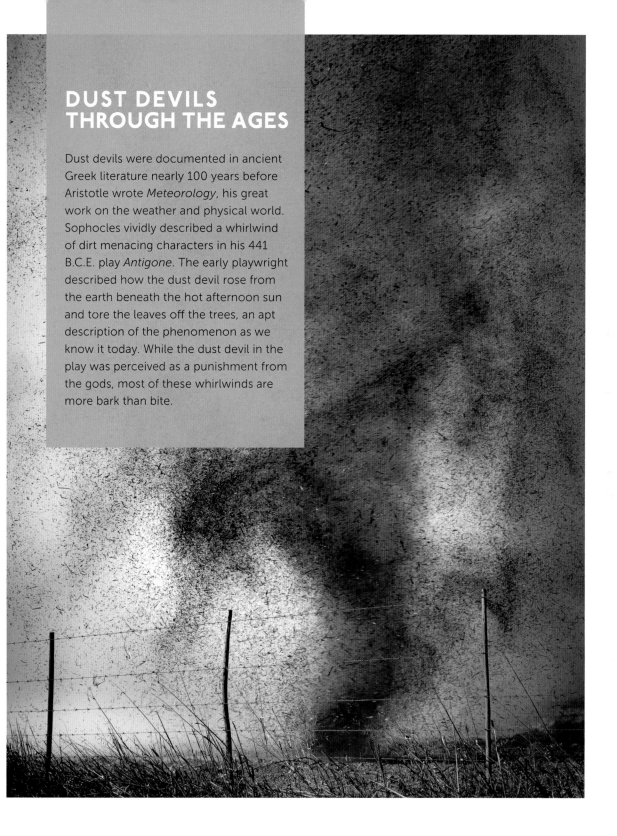

# DUST DEVILS THROUGH THE AGES

Dust devils were documented in ancient Greek literature nearly 100 years before Aristotle wrote *Meteorology*, his great work on the weather and physical world. Sophocles vividly described a whirlwind of dirt menacing characters in his 441 B.C.E. play *Antigone*. The early playwright described how the dust devil rose from the earth beneath the hot afternoon sun and tore the leaves off the trees, an apt description of the phenomenon as we know it today. While the dust devil in the play was perceived as a punishment from the gods, most of these whirlwinds are more bark than bite.

heat generated by a raging wildfire rises into the atmosphere and creates its own weather conditions. Air rushes into the fire to replace the enormous amount of air rising skyward, creating powerful winds that grow and spread the flames.

The combination of intense winds rushing into the fire and an intense updraft forming above the fire can trigger fire whirls. These whirlwinds form in a similar manner to dust devils, just on a much larger and more extreme scale. A fire whirl can grow so large and powerful that the swirling flames and smoke are easily detected by Doppler weather radar. Meteorologists in Northern California even issued a tornado warning for a fire whirl in July 2018. They sent a crew to survey the damage afterward and found that the fire whirl produced damage equivalent to an EF-3 tornado, with estimated winds greater than 140 miles per hour.

# RAINBOWS

**E**very thunderstorm and torrential downpour eventually comes to an end. And sometimes, the Sun comes out and treats a drenched community to a spectacular sight. A brilliant rainbow glistening against the backdrop of a departing thunderstorm means you found yourself in the right place at the exact right time to see this lovely optical show.

## Prisms

A rainbow starts with the speed of light, which travels at more than 180,000 miles per second when it's in the vacuum of space. Here on Earth, the speed of light varies a little bit depending on how much stuff the light has to go through before it reaches your eyes. Light slows down when it passes through gases, liquids, and solids that have different densities. When light suddenly slows down as it passes from lower density air to higher density water, for instance, the light bends slightly. This bend is called refraction.

A simple glass of water provides a great example of refraction. If you place a straw in the glass and look at it from the side, the straw appears "broken" or displaced beneath the water line. This is the result of light refracting as it passes through the air and into the water, then back into the air on its way to your eye.

The effect is often demonstrated using a glass prism, or a small piece of triangular glass. Light slows down and bends as it moves through the air and into the glass prism.

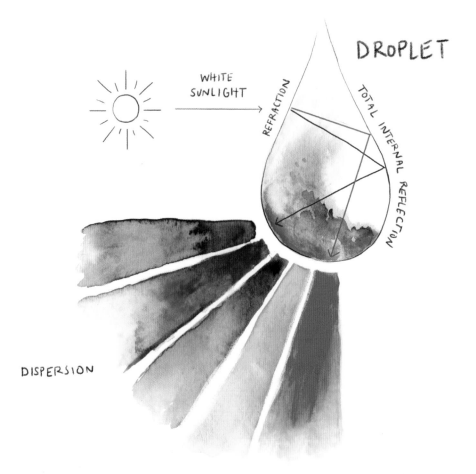

WHITE SUNLIGHT

DROPLET

REFRACTION

TOTAL INTERNAL REFLECTION

DISPERSION

This visible light is made up of different wavelengths. Our eyes see each wavelength as a different color. Red makes up the longest wavelength in visible light, while violet makes up the shortest wavelength.

When light bends through the prism, the visible light disperses into its different colors. Longer wavelengths bend the least as they pass through the prism because they're moving the fastest, while shorter and slower wavelengths spend more time in the glass and bend the most. Red light leaves the prism first while violet light leaves the prism last. The result is visible light that's split into its component colors: red, orange, yellow, green, blue, indigo, and violet—all the colors of the rainbow.

## RAINLESS RAINBOWS

If you crave a rainbow and you just can't wait for the next perfectly timed rainstorm, you're in luck. Rainbows are one of the few weather events we can create with little effort. Lawn sprinklers are great for creating your own personal rainbow, especially if the water comes out of the nozzle in a heavy sheet or a fine mist. You can even use a spray bottle to recreate the effect on demand.

## Shiny Raindrops

Raindrops falling through the sky can act like tiny prisms that refract and disperse light back to your eye. But it's a complicated process. Not all refraction results in a rainbow. After all, every glass of water doesn't light up the world like a nightclub. It's all about the angles. The sunlight, your eyes, and the water droplets in the air have to line up in just the right order—the Sun behind us and rain falling in front of us—for you to see a rainbow.

Raindrops produce a rainbow by reflecting light, refracting light, and dispersing light. Imagine you're staring at a downpour to the east while the Sun approaches sunset in the west. The sunlight emanating from behind you shines into the raindrops in front of you. The light slows down as it enters the water droplet because water has a higher density than air, which causes the light to start bending. This refracted sunlight then reflects off the back of the raindrop like a mirror, redirecting the sunlight back toward your eye. Refraction causes the visible light to disperse, allowing each color in the

# MOONBOWS

A clear night with bright moonlight can create a moonbow. Moonbows and rainbows are more like cousins than siblings. Since moonlight is just sunlight reflected off the lunar surface, the light of a full moon isn't nearly as strong as direct sunlight. As a result, moonbows are relatively rare and relatively faint. They're so faint, in fact, that they usually appear as a white arc rather than a colorful spectrum.

visible light spectrum to bend at a slightly different angle. This splits the light into each color that we see as a rainbow.

## It's All in the Angle

A well-formed rainbow appears as a brilliant arch in the sky because sunlight has to exit the raindrops at an angle of 42° to reach your eye as a vivid display of colors. If you held up a protractor to the sky, the sky measures 180° from the horizon in front of you to the horizon behind you, with 90° representing the sky straight above your head.

Since sunlight has to leave the raindrop at 42° in order to reach your eye as an array of pretty colors, the rainbow always appears to be 42° in the sky from the right, from the left, and from above. That's why a classic rainbow always appears roughly the same size, shape, and distance no matter where you stand. This is also why each rainbow is unique to you. Since it takes countless raindrops to create a kaleidoscope of colors, the raindrops that produced your rainbow aren't the same raindrops that produce the rainbow viewed by the person next to you, or even by the camera in your hand.

The appearance of a rainbow depends on the intensity of the sunshine behind you and the heaviness of the rainfall in front of you. These formations can change from one second to the next, depending on how fast the clouds and rain are moving, so a rainbow may disappear quickly.

The ideal, picturesque rainbow forms when the sky is perfectly clear toward the Sun and there's tons of rain ahead of you. Spotty rainfall or clouds obstructing the sunlight can lead to faint colors that appear to only cover a tiny portion of the sky, most often near the horizon. If the Sun is too high in the sky, if there are too many clouds blocking the Sun, or if the raindrops are too few and far between, you won't see a rainbow at all.

If the sunlight hits the raindrop at just the right angle, the light can reflect twice within each single raindrop and produce a double rainbow. When you have the good fortune to witness these twin formations, a fainter and larger secondary rainbow forms above the primary rainbow below.

Our skies are never boring. Each snow shower and heat wave tells its own story. Clouds that hover over mountaintops and loom beneath thunderstorms form through processes that meld intense forces with delicate precision. Every fleeting rainbow can seem commonplace until you learn that each version is unique to you.

My love for the weather began as a child. I grew up near Washington, DC, where you get to sample just about every type of extreme weather imaginable. Big blizzards and raging thunderstorms can leave a lasting mark on an impressionable little kid. I still feel that same sense of awe when I gaze at lightning flashing in a distant thunderstorm or catch a bright meteor zipping across the sky.

That kind of inspiration is what I hoped to accomplish with *The Skies Above*. Whether you've loved storms since you were a kid or you never really noticed the weather beyond checking if you need an umbrella, my sincere hope is that you'll take a moment to appreciate the calm and recognize the power behind the calamity.

Even if you don't fall head over heels in love with every raindrop or storm system that rolls through your area, I hope that you've taken in enough of the basics that you're not caught off-guard when the skies darken, and that knowledge gives you some power over the fear. Weather forecasting is one of humankind's greatest achievements, and keeping up with approaching storms is one of the easiest ways we can keep ourselves safe.

If we can learn anything from admiring the sky above us, it's that the weather is never just small talk. Even the sunniest day is teeming with wonder and possibility.

RESOURCES

**Climate Prediction Center**: The CPC is the place to go if you're looking for information on weather patterns in the coming weeks and months, as well as the latest status on El Niño or La Niña. (cpc.ncep.noaa.gov)

**COMET MetEd**: Hundreds of free, college-quality lessons on just about every meteorology and climatology topic you might want to learn about, operated by the COMET Program within the University Corporation for Atmospheric Research (UCAR). (meted.ucar.edu)

***Essentials of Meteorology***: The quintessential introduction to meteorology textbook from C. Donald Ahrens is a wonderful gift for a budding meteorology student or for everyday reference on your bookshelf.

**Iowa Environmental Mesonet**: The IEM is a treasure trove for past weather conditions and data visualization for just about every weather station in the world. (mesonet.agron.iastate.edu)

**mPing**: A handy mobile app that lets you report precipitation and severe weather to meteorologists right from your phone. This information helps forecasters issue warnings and verify radar data. (mping.nssl.noaa.gov)

**National Hurricane Center**: A visit to the NHC should be a daily routine during hurricane season, as the agency is responsible for predicting and tracking tropical activity in the Atlantic Ocean and the eastern and central Pacific Ocean. (nhc.noaa.gov)

**National Weather Service**: The US government's official forecasting agency, the 122 offices of the NWS are a fantastic local resource for accurate forecasts and current weather data. (nws.noaa.gov)

**NOAA Weather Radio**: Weather radios are like smoke detectors for dangerous storms. You can program these radios to sound a loud alarm when severe weather watches or warnings are issued for your area. (www.weather.gov/nwr)

**NWS SKYWARN**: The NWS's official storm spotter program offers a bevy of free courses around the country each year, training everyday citizens to spot and report hazardous weather. (weather.gov/SKYWARN)

**RadarScope**: If you want to stay one step ahead of rain and storms heading your way, RadarScope is a must-have radar app for mobile devices and computers alike. (radarscope.app)

**Storm Prediction Center**: A branch of the NWS, this agency is an invaluable wealth of information during severe weather season. (spc.noaa.gov)

**The Community Collaborative Rain, Hail, and Snow Network**: CoCoRaHS, as it's called, is a group of more than 20,000 volunteers in the US and Canada who collect and share daily precipitation measurements. (cocorahs.org)

***Weather: A Golden Guide***: A timeless classic from St. Martin's Press, this pocket-size book shares succinct explanations of common weather topics.

***Weather Analysis & Forecasting Handbook***: This popular handbook by meteorologist Tim Vasquez provides a solid foundation in the basic principles of meteorology and forecasting the weather.

**WeatherBrains**: This long-running podcast by a group of entertaining meteorologists shares insights, history, and riveting conversations from interviews with hundreds of guests. (weatherbrains.com)

**Weather Prediction Center**: The WPC is a fantastic resource for surface weather maps, as well as snow, ice, and heavy rainfall forecasts. (wpc.ncep.noaa.gov)

## Image Credits

**Alamy Stock**: Media Drum World, 133; **Jeremy Perez**, 215; **Little Firefly Photography**, 231; **NASA**: 56, 34; NASA/JPL-Caltech/MSSS, 173; **Rebecca Matt**: 11, 16, 28, 48, 49, 62, 69, 80, 93, 96, 121, 157, 188, 201, 220; **Shutterstock.com**: 24Novembers, 72; Alex Stemmers, 78, 79; Alexander_P, 154; Andrei Stepanov, 87; Anna in Sweden, 83; Anton Petrus, 156, 157; ArtMari, 132; Artskrin, 198; Artur Balytskyi, 74, 108, 150, 185; Bodor Tivadar, 82, 138, 221; cdstocks, 151; CE Photography, 31; ch123, 25; Christian Roberts-Olsen, 45; Christian Schwier, 123; Creative Travel Projects, 112, 113; Curioso.Photography, 163; Diane C Macdonald, 31; Diego Rebello, 66, 67; Digital signal, 109; Elala, 135; elcatso, 186; Everett Collection, 170, 171; Fer Gregory, 130; Gail Johnson, 177; Glynnis Jones, 202, 203; GUAVWA, 140; Holiday62, 148; Ian Lienhard, 107; Jake Hukee, 94, 95; James Mattil, 219; Jaromir Chalabala, 6, 7, 18, 19; Jeffrey M. Frank, 211; jo Crebbin, 183; John D Sirlin, 118, 119, 166, 167, 169; Kate Macate, 89; Kay Cee Lens and Footages, 30; Kseniakrop, 125; Lisa DeShantz-Cook, 10, 11; Lisovskaya Oksana, 128; logoboom, 144, 145; lynea, 120; Maksim Shmeljov, 184; Menno van der Haven, 29, 105; Michael Dorogovich, 34; Minerva Studio, 212, 213; Mitch Boeck, 98; Morphart Creation, 46, 116, 152; Nadiinko, 24, 97; NASA images, 38; Nicky Elliott, 217; Nikolayenko Yekaterina, 57; Octavian Cocolos, 60, 61; Petr Malyshev, 31; Phattaraphum, 29; PhilMacDPhoto, 137; Piyaset, 43; Ray_Bemish, 41; Shaiith, 100, 101; Siraphob Werakijpanich, 51; Stefano Garau, 30; Steve Cukrov, 59; Steve Photography, 84; Sunny Forest, 27; Svetoslav Iliev, 35; THANAKRIT SANTIKUNAPORN, 52, 53; Triff, 194, 195; Trong Nguyen, 22; TTstudio, 180, 181; turtix, 75; VanderWolf Images, 32, 33; Various images, 63; VectorMine, 164; Vitezslav Halamka, 68; Vladimir Melnik, 117; Wirestock Creators, 207; Worachat Limleartworakit, 174, 175; Yarygin, 90; Yaya Ernst, 224; YummyBuum, 14; **Wikimedia**: Arne-Kaiser, 222; Senior Airman Brian Kelly, 126; U.S. Air Force/Staff Sgt. Manuel J. Martinez, 205; Zombiepedia, 65;

Little Firefly Photography

Dennis Mersereau enjoys all types of atmospheric tantrums, from powerful thunderstorms to heavy snow. After studying meteorology in college, he went off to write for the *Washington Post*'s Capital Weather Gang and Gawker's weather vertical, The Vane, as well as for *Popular Science, Mental Floss, Forbes,* and The Weather Network. He also teamed up with survival experts to write *The Extreme Weather Survival Manual* (Outdoor Life, 2015). He lives in North Carolina.

**MOUNTAINEERS BOOKS**

recreation • lifestyle • conservation

**MOUNTAINEERS BOOKS** is a leading publisher of mountaineering literature and guides—including our flagship title, *Mountaineering: The Freedom of the Hills*—as well as adventure narratives, natural history, and general outdoor recreation. Through our two imprints, Skipstone and Braided River, we also publish titles on sustainability and conservation. We are committed to supporting the environmental and educational goals of our organization by providing expert information on human-powered adventure, sustainable practices at home and on the trail, and preservation of wilderness.

The Mountaineers, founded in 1906, is a 501(c)(3) nonprofit outdoor recreation and conservation organization whose mission is to enrich lives and communities by helping people "explore, conserve, learn about, and enjoy the lands and waters of the Pacific Northwest and beyond." One of the largest such organizations in the United States, it sponsors classes and year-round outdoor activities throughout the Pacific Northwest, including climbing, hiking, backcountry skiing, snowshoeing, camping, kayaking, sailing, and more. The Mountaineers also supports its mission through its publishing division, Mountaineers Books, and promotes environmental education and citizen engagement. For more information, visit The Mountaineers Program Center, 7700 Sand Point Way NE, Seattle, WA 98115-3996; phone 206-521-6001; www.mountaineers.org; or email info@mountaineers.org.

Our publications are made possible through the generosity of donors and through sales of 700 titles on outdoor recreation, sustainable lifestyle, and conservation. To donate, purchase books, or learn more, visit us online:

**MOUNTAINEERS BOOKS**
1001 SW Klickitat Way, Suite 201 • Seattle, WA 98134
800-553-4453 • mbooks@mountaineersbooks.org • www.mountaineersbooks.org

*An independent nonprofit publisher since 1960*

YOU MAY ALSO LIKE

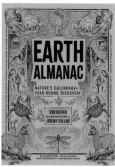
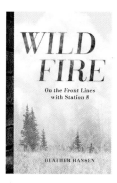